Cryo-EM from Cells to Molecules: Multi-Scale Visualization of Biological Systems 2018

Tahoe City, California, USA
4 - 8 February 2018

ISBN: 978-1-5108-6255-5

Printed from e-media with permission by:

Curran Associates, Inc.
57 Morehouse Lane
Red Hook, NY 12571

Some format issues inherent in the e-media version may also appear in this print version.

Printed by Curran Associates, Inc. (2018)

For permission requests, please contact Keystone Symposia
at the address below.

Keystone Symposia
PO Box 1630
Silverthorne, CO 80498
USA

Phone: +1 800-253-0685 or +1 970-262-1230
Fax: +1 970-262-1525

info@keystonesymposia.org

Additional copies of this publication are available from:

Curran Associates, Inc.
57 Morehouse Lane
Red Hook, NY 12571 USA
Phone: 845-758-0400
Fax: 845-758-2633
Email: curran@proceedings.com
Web: www.proceedings.com

Cryo-EM from Cells to Molecules: Multi-Scale Visualization of Biological Systems

Scientific Organizers:
Georgios Skiniotis
Elizabeth Villa
Andrew B. Ward

Sponsored by:
Pfizer Inc.

February 4–8, 2018
Granlibakken Tahoe
Tahoe City, California, USA

#KScryoEM

Visit **www.keystonesymposia.org/18F1** to view the conference program online.

Unless otherwise noted, the information in this book is current as of **January 15, 2018**. If you registered after this date, your name is included in an online list accessed from attendees'

Please be advised that no video equipment, cameras, audio equipment or any other type of recording device will be allowed in the conference room or poster sessions

KEYSTONE SYMPOSIA™
on Molecular and Cellular Biology
Accelerating Life Science Discovery

Keystone Symposia is a 501(c)(3) nonprofit organization directed and supported by the scientific community.
info@keystonesymposia.org | www.keystonesymposia.org

Keystone Symposia Policies

Harassment Policy

Keystone Symposia is committed to maintaining a positive and respectful environment at its conferences and other events. We expect participants in our events to engage in constructive and professional discussion, in which all are valued for their scientific contributions and work. We value diversity, and desire that no participant should be subjected to harassment while involved in our events.

For purposes of this policy, harassment means unwelcome and offensive comments or behavior directed to the participant's sex, race, color, national origin, religion, sexual orientation or gender identity, disability or other status protected under applicable law. Harassment can include, for example, unwelcome attention, comments or jokes that focus on gender differences or sexual topics and that distract from the professional topics under discussion, unwelcome advances or requests for dates or sexual activities, and the use of language or images that demean or degrade persons of particular gender, racial, ethnic, religious or national identity.

To this end, we expect all participants to support these values and to avoid harassment of others participating in our conferences and other events. We expect all attendees to assist in ensuring that Keystone Symposia events are free from harassment of any kind, including reporting any instances of harassment directly to Dr. Jane Peterson, CEO, at **janep@keystonesymposia.org** and/or Dr. Thale Jarvis, CSO, at **thalej@keystonesymposia.org**. Anyone who has experienced harassment, or who has witnessed such behavior, should notify one of the above persons as soon as possible.

Persons who act contrary to these values and expectations may be warned or asked to leave the event in which the behavior occurred, may be excluded from access to Keystone Symposia conferences and/or other events, and/or may be subject to other disciplinary or corrective action, at the discretion of Keystone Symposia.

Privacy Policy

Keystone Symposia is committed to protecting the privacy of its website visitors and meeting attendees. Keystone Symposia collects personal information when individuals register for our meetings and upon a visitor's request to subscribe to newsletters and meeting announcements (both print and online). Information that our visitors and attendees provide or that is derived from internal website tracking is not sold, rented or shared with any third-party individual or organization. Once on our mailing lists, individuals always have the option of unsubscribing so that they no longer receive all or certain types of our communications.

By participating in a Keystone Symposia meeting, attendees acknowledge that their name and photograph may be published in a limited fashion in materials produced by Keystone Symposia. For example, to make the meeting a more valuable experience for all involved, attendee names and institutions are listed on our website in a secure section accessible only by attendees of the same meeting. Attendee names and institutions are also pre-populated in our secure mobile app, but attendees decide for themselves whether to enter further information in the app.

Attendee names and contact information are also listed in the meeting book. Except in the meeting book in this fashion, we will not disclose attendee contact information, even to other attendees. Photographs of meeting interactions taken by Keystone Symposia may occasionally be used in our marketing literature.

For our full privacy policy, please visit **www.keystonesymposia.org/privacy**.

If you have any further questions or comments about our privacy policy, please send your feedback to **info@keystonesymposia.org**.

Meeting Room Policy

No video equipment, cameras, audio equipment or any other type of recording device will be allowed in the meeting rooms or poster sessions. Occasionally, Keystone Symposia may use such devices for its own publicity purposes. While we do not prohibit laptop computers, cell phones and PDAs, they must not be used for recording and should be operated in "silent" mode out of consideration for speakers and other conference attendees.

Attendees will be required to wear name badges for access to meeting sessions. Due to problems we have encountered at some meetings, we will be performing random badge checks. Attendees without badges will be turned away from the session.

Please note that spouses traveling with registered conference attendees are not permitted inside the sessions unless they pay the registration fee. However, they are welcome to attend evening receptions. They may attend breakfasts upon payment of a nominal daily fee.

Media and Communications Policy

Keystone Symposia recognizes that presenters of scientific data may have reasons for not wanting early results reported to the general public prior to peer review. We also recognize, however, that raising society's level of science knowledge and awareness is essential for appropriate scientific input into public policy and decision-making by political leaders, which is in everyone's best interest. We therefore encourage and will try to facilitate interactions between the scientists attending our conferences and the media. We ask both to be understanding when considering each other's objectives and the overarching goal of raising science literacy worldwide.

If approached with sufficient advance notice, Keystone Symposia can provide assistance to journalists to contact our speakers and abstract authors directly. We can prearrange interviews with specific meeting organizers, speakers, authors or Keystone Symposia staff.

We ask that all writers attending a Keystone Symposia conference gain approval from a speaker or poster presenter prior to quoting or publishing that individual's scientific results. This policy applies whether you are a professional writer/journalist or a non-journalist blogging about the conference or otherwise sharing information among a group of individuals.

Audio, still photo and video recording by any device (e.g., cameras, laptops, PDAs, cell phones, watches) is strictly prohibited during the sessions, unless in certain circumstances when prior permission must be obtained from Keystone Symposia. Photographs taken by Keystone Symposia may be available on request.

Keystone Symposia welcomes members of the scientific and general media at our meetings. Due to the costs of providing meals and other facilities, payment of the regular registration fee is required. Keystone Symposia may be able to give some consideration to journalists from nonprofit organizations, as well as to journalists who wish to attend the meeting for just one day. Such inquiries and arrangements should be made in advance.

Keystone Symposia does not require our presenters to submit papers, nor do we record or transcribe our sessions. Speaker abstracts are available in the meeting book provided to each registrant. Meeting books are available after the meeting to non-attendees for a nominal fee.

Keystone Symposia provides a venue for scientists to come together and share their ideas with each other in a relaxed setting. While we wish to accommodate members of the press, we ask that all members of the media respect our mission and the freedom we allow our scientists to discuss their work in a protected and informal environment.

If you have questions about this policy, contact Yvonne Psaila, Director of Marketing and Communications, **yvonnep@keystonesymposia.org**, **1.970.262.2676**.

Policy for Displaying Literature at Keystone Symposia Meetings

Keystone Symposia provides financial and in-kind donors with the ability to display a limited amount of literature or small promotional items on a table near the registration desk at Keystone Symposia meetings. Financial donors at certain levels can also insert literature into delegate bags at the specific meeting(s) they have chosen to support. Literature placed by organizations that are not current supporters will be removed until the meeting participant who placed it can be contacted.

Please contact the following Keystone Symposia staff if you are interested in displaying literature at our meetings and becoming a donor:

Financial support (e.g., corporations, foundations):	**Sarah Lavicka**, Director of Development **1.970.262.2690; sarahl@keystonesymposia.org**
In-kind support (e.g., publishers, societies/associations, conference organizers):	**Yvonne Psaila**, Director of Marketing and Communications **1.970.262.2676; yvonnep@keystonesymposia.org**

Nonprofit, academic and government institutions may post a flyer about grant and job opportunities or upcoming events of interest on the bulletin board in the registration area, and job postings can also be placed on the meeting mobile app. Please see a Keystone Symposia on-site representative if you need help with this or have questions while at the conference.

Welcome

Dear Conference Attendee,

Welcome to the Keystone Symposia 2017-2018 series and thank you for attending this conference. We are confident you will benefit from the face-to-face contact, opportunities for collaboration-building, deep learning and respectful debate that occur in the intimate environment of our conferences.

Two years ago, the process began when we worked with our Scientific Advisory Board to program this conference and select its volunteer scientist organizers, with whom we then work closely to compile as diverse and interesting a speaker roster as possible and to craft a program that reflects the latest science. To provide opportunities for early-career investigators and late-breaking research, short talk slots are left open until three months before the meetings and chosen based on abstracts submitted by the abstract deadline.

We have 60 conferences in our 2018 season from September 2017 to June 2018. They span all areas of the life sciences and include topics we have not held before such as Maternal-Fetal Crosstalk, Novel Aspects of Bone Biology, Organs- and Tissues-on-Chips, and Frontiers of Serotonin Beyond the Brain. They will take place in eight countries: Canada, Germany, Hong Kong, Japan, South Africa, Sweden, the UK and the US. If this is your first Keystone Symposia conference, we hope you will come back for future meetings and urge your colleagues to do the same. In between conferences, please take advantage of the digital content we have assembled at **virtual.keystonesymposia.org**, much of which is free.

We thank the organizers and all others involved in putting this conference together as well as everyone participating. We hope that you enjoy it. You will receive a link to an electronic survey as the conference concludes, but as always, please don't hesitate to send any feedback directly to **info@keystonesymposia.org** or **janep@keystonesymposia.org**.

Sincerely,

Jane L. Peterson, PhD
President and Chief Executive Officer
Keystone Symposia

Gary J. Nabel, MD, PhD
Chief Scientific Officer, Sanofi
Chair of the Board, Keystone Symposia

Maximimizing Your Experience

The conference program has been designed around our mission: to accelerate life science discovery by providing a forum to present top-quality science, foster new collaborations and help prepare the next generation of life scientists.

Poster Abstract Sessions

Poster sessions play host to some of the most dynamic interactions that take place at our conferences and are not to be missed. Abstracts are numbered by session: abstracts presented during Poster Session 1 are numbered in the 1000s, Poster Session 2 in the 2000s, etc. If you are a presenter, please check the index to find your abstract number; poster abstracts can be viewed on your Keystone Symposia online account and on the dedicated meeting mobile app.

Meals

The meals included in your registration vary by site. Check the program for meals marked "On Your Own" and plan accordingly. Meals listed with a time and place are provided as part of your registration. Some attendees choose to make a meal out of the evening social hour. Please note that alcohol and entertainment are not funded by registration fees or US government grants. Funding for this expense is generously provided by other supporters of Keystone Symposia.

Digital Meeting Book from Your Account

You can download a PDF of the meeting book from your Account on our secure website beginning seven days before the meeting and for up to 90 days afterwards. Your Account page also contains other useful content such as printable invoices and invitation letters, your profile with mail/email preferences and much more. Speaker and poster abstracts are also available here, as well as on the meeting mobile app.

We Want Your Feedback!

Please be sure to give us your feedback by completing the survey that will be emailed to you at the conclusion of the meeting, as well as the briefer one on the mobile app. Your input is very valuable to us as we plan future conferences.

Cryo-EM from Cells to Molecules: Multi-Scale Visualization of Biological Systems

Scientific Organizers:
Georgios Skiniotis
Elizabeth Villa
Andrew B. Ward

Sponsored by:
Pfizer Inc.

February 4–8, 2018
Granlibakken Tahoe
Tahoe City, California, USA

Fast-evolving cryo-EM visualization techniques have recently started to illuminate the inner workings of complicated and dynamic biological assemblies with a level of detail we have never seen before. Cryo-EM is now the state-of-the-art methodology for taking snapshots of biological systems from low to high resolution and with various levels of complexities, from small macromolecular complexes all the way to subcellular organizations. This symposium will showcase cryo-EM breakthroughs in addressing crucial problems in cell biology and will further highlight cryo-EM imaging as a platform for the integration of data from orthogonal approaches. With talks on a variety of biological specimens, procedures and types of biological questions, the meeting aims to attract a very diverse audience from both academia and industry with interest in the method, and beyond the traditional cryo-EM specialist. The symposium will focus on four themes: 1) Single-particle cryo-EM with attention to the characterization of the conformational ensemble adopted by macromolecular complexes and aiming to delineate their mechanism of action; 2) Cryo-electron tomography (cryo-ET) and correlative light electron microscopy (CLEM) to visualize cellular organizations and their dynamics; 3) The emergence of cryo-EM as tool for drug discovery and translational research through both high-resolution structure determination and characterization of cellular phenotypes; and 4) Recent developments in sample preparation and image acquisition. Furthermore, the symposium will include workshops in specimen preparation and evaluation of suitability for cryo-EM, modeling of structures using cryo-EM data, and how to set up a cryo-EM lab.

Schedule may have changed since this book was prepared. Visit ***www.keystonesymposia.org/18F1*** *and the conference mobile app at* ***apps.eventmobi.com/ks*** *for the most up-to-date program and for all speaker and poster abstracts. Hashtag for this meeting:* ***#KScryoEM***

SUNDAY, FEBRUARY 4

16:00–20:00	Pre Function	Arrival and Registration
18:00–20:00	Granhall	Welcome Mixer
18:00–20:00	Granhall	Light Meal – Provided by Granlibakken

MONDAY, FEBRUARY 5

07:00–08:00	Granhall	Breakfast
08:00–09:00	**Mountain/Lake**	**Welcome and Keynote Address** ***Georgios Skiniotis**, Stanford University, USA ***Andrew B. Ward**, The Scripps Research Institute, USA ***Elizabeth Villa**, University of California, San Diego, USA **Patrick Cramer**, Max Planck Institute for Biophysical Chemistry, Germany *Integrated Structural Biology of Gene Transcription.....2*
09:00–11:30	**Mountain/Lake**	**Visualizing Cells** ***Elizabeth Villa**, University of California, San Diego, USA *Opening Windows into the Cell:* *Bringing Structure to Cell Biology using Cryo-Electron Tomography.....N/A* Coffee Break **Grant J. Jensen**, California Institute of Technology, USA *Structure and Function of the Bacterial Type IV Secretion System through Electron Cryotomography.....5* **Mark H. Ellisman**, University of California, San Diego, USA *Revealing Secrets Hiding in Plain Sight.....3* **Christina Heroven**, MRC Laboratory of Molecular Biology, UK (1031) *Short Talk: In-situ Structural Studies of RPTPmu by Cryo-Electron Tomography of FIB Milled Cells.....25* **Kanika Khanna**, University of California, San Diego, USA (2008) *Short Talk: Understanding Morphological Changes Accompanying Sporulation in B. subtilis using in situ Cryo-Electron Tomography......30* On Own for Lunch
11:30–13:00	Bay	Poster Setup
13:00–22:00	Bay	Poster Viewing
16:30–17:00	Pre Function	Coffee Available

All abstracts are now available on your Keystone Symposia online account or our meeting mobile app at ***apps.eventmobi.com/ks****. Full talk speaker abstracts are in the "Speaker Abstracts" section, and short talk speaker abstracts are in the "Poster Abstracts" section (poster abstract number is shown in parentheses in the program above).*

MONDAY, FEBRUARY 5 (continued)

17:00–19:00	**Mountain/Lake**	**Visualizing Subcellular Compartments** **Daniela Nicastro**, University of Texas Southwestern Medical Center, USA *Probing the Molecular Organization of Cells and Organelles using Cryo-Electron Microscopy.....7* **Jun Liu**, Yale School of Medicine, USA *High-Throughput Cryo-Electron Tomography: Visualizing Host-Pathogen Interactions at High Resolution.....5* ***Kay Grünewald**, University of Oxford, UK *Trafficking Vesicles and Enveloped Viruses* **Joshua Hutchings**, Birkbeck College, UK (1032) *Short Talk: Cryo-EM Tomography and Subtomogram Averaging of COPII Assembled on Membranes.....26*
19:00–20:00	Granhall	Dinner
19:30–22:00	**Bay**	**Poster Session 1** *Poster sessions provide exciting opportunities for engagement between all levels of investigators. Abstracts beginning with the number 1 are featured during this poster session. Don't miss the plasma screen e-posters during this session!*
20:00–21:00	Bay	Social Hour

TUESDAY, FEBRUARY 6

07:00–08:00	Granhall	Breakfast
08:00–11:15	**Mountain/Lake**	**Translational Cryo-EM** ***Andrew B. Ward**, The Scripps Research Institute, USA *Structure-Based Vaccine Design.....8* **Tom Ceska**, UCB Celltech, UK *Cryo EM as a Tool for Structure Determination in Industry.....2* Coffee Break **Seungil Han**, Pfizer Inc., USA *Applications of Cryo-EM in Small Molecule and Biologics Drug Design.....4* **Christian Wiesmann**, Novartis, Switzerland *Cryo-Em: A New Tool for Drug Discovery at Novartis.....9* **Gregory Martin**, Oregon Health & Science University, USA (2024) *Short Talk: Anti-Diabetic Drug Binding Site in KATP Channels Revealed by Cryo-EM.....38* **Chitra Shintre**, University of Oxford, UK (3015) *Short Talk: Structure of the Polycistic Kidney Disease TRP Channel PC2.....48* On Own for Lunch
11:15–13:00	Bay	Poster Setup
13:00–22:00	Bay	Poster Viewing

*Session Chair; ( ) = abstract number

TUESDAY, FEBRUARY 6 (continued)

14:30–16:30	**Mountain/Lake**	**Workshop 1: Specimen Preparation and Troubleshooting for Single-Particle Work** ***Iban Ubarretxena**, Icahn School of Medicine at Mount Sinai, USA (3032) *Structures of Key Intermediates in the Human Mitochondrial Chaperonin Reaction Cycle.....53* **Anika L. Burrell**, University of Washington, USA (1012) *Human IMPDH1 Retinal Degeneration Mutations Promote Polymer Assembly.....15* **Joel Butterwick**, Rockefeller University, USA (1013) *Cryo-EM Structure of Orco: The Molecular Basis for Insect Olfaction.....16* **Ziao Fu**, Columbia University, USA (1027) *Lipid Bilayer Structure in Native Cell Membrane Nanoparticles of Multidrug Exporter AcrB.....23* **Jonas Barandun**, Rockefeller University, USA (1004) *Structure of the Earliest Stable Precursor of the Ribosome – the Small Subunit Processome.....11* **Vignesh Kasinath**, University of California, Berkeley, USA (2006) *Cryo-EM Structure of the Full Human PRC2 and Regulation by AEBP2 and JARID2.....29* **Qianhui Qu**, Stanford University, USA (3003) *Structure of a Histone H3K4 Methyltransferase COMPASS.....43* **Julia Brasch**, Columbia University, USA (1009) *Assembly of Clustered Protocadherin Neuronal Recognition Complexes Studied by X-ray Crystallography and Cryo-electron Tomography.....14* **Cristina Puchades**, The Scripps Research Institute, USA (3002) *Hand-Over-Hand: How ATP Drives Protein Translocation.....42* **Harry Low**, Imperial College, UK (2017) *Architecture of a Bacterial Type II Secretion System.....35*
16:30–17:00	Pre Function	Coffee Available
17:00–19:00	**Mountain/Lake**	**Frontiers in Cryo-EM** **Andrej Sali**, University of California, San Francisco, USA *Integrative Modeling of Biomolecular Assembly Structures and Pathways.....8* ***Sharon Grayer Wolf**, Weizmann Institute of Science, Israel *Deep Views into Vitrified Cells by Cryo-Scanning Transmission Electron Tomography.....9* **David Baker**, University of Washington, USA *De novo Design of Self-Assembling Protein Nano Materials .....1* **Yuxi Liu**, University of California, Los Angeles, USA (2015) *Short Talk: Near-Atomic Resolution Cryo-EM for a Small Protein on a Symmetric Imaging Scaffold......34*
19:00–20:00	Granhall	Dinner
19:30–22:00	**Bay**	**Poster Session 2** *Poster sessions provide exciting opportunities for engagement between all levels of investigators. Abstracts beginning with the number 2 are featured during this poster session. Don't miss the plasma screen e-posters during this session!*
20:00–21:00	Bay	Social Hour

*Session Chair; ( ) = abstract number

WEDNESDAY, FEBRUARY 7

07:00–08:00	Granhall	Breakfast
08:00–11:00	**Mountain/Lake**	**Cryo-EM of Small Single Particles** **Dmitry Lyumkis**, The Salk Institute for Biological Studies, USA *High-Resolution Cryo-EM Studies of Retroviral Intasomes.....6* **Maofu Liao**, Harvard Medical School, USA *Single Particle Cryo-EM Studies of Lipid-Protein Interactions.....5* Coffee Break ***Gabriel C. Lander**, The Scripps Research Institute, USA *How Low Can You Go? Size and Resolution Limits using Conventional Cryo-EM.....5* **Moran Shalev-Benami**, Stanford University, USA (3014) *Short Talk: Cryo-EM Structure of the Synaptic Wiring Protein Teneurin.....47* **Christopher P. Hill**, University of Utah, USA (2033) *Short Talk: Structural Basis for Protein Translocation by the Vps4 AAA ATPase.....25* **Daniel Southworth**, University of California, San Francisco, USA (3036) *Short Talk: Visualizing the Substrate Translocation Mechanism of the Hsp104 AAA+ Disaggregase.....48* On Own for Lunch
11:00–13:00	Bay	Poster Setup
13:00–22:00	Bay	Poster Viewing
14:30–16:30	**Mountain/Lake**	**Workshop 2: Modeling and Refining Structures of Macromolecules in Cryo-EM Maps** ***Gunnar Schroeder**, Forschungszentrum Julich, Germany (3013) *Fibril Structure of Amyloid-β(1-42) by Cryo-Electron Microscopy.....47* **Jane S. Richardson**, Duke University, USA (3007) *CaBLAM – A New Tool Designed to Validate 2.5-4Å Models from Cryo-EM* **Doo Nam Kim**, Los Alamos National Laboratory, USA (2010) *CryoFIT: Flexible Fitting of Biomolecule Structures from Cryo-EM Maps with User-Friendly Molecular Dynamics Simulation* **Jianpeng Ma**, Baylor College of Medicine, USA (2021) *Application of Parallel Continuous Simulated Tempering to Cryo-EM Structural Refinement at Near-Atomic Resolution* **Abhishek Singharoy**, Arizona State University, USA (3016) *Molecular Dynamics Flexible Fitting at sub-5 Å Resolution* **Till Rudack**, Ruhr University Bochum, Germany (3010) *ModelMaker: An Interactive Integrative Modeling Tool to Build Complete Macromolecular Machines* **Kelly Brock**, Harvard Medical School, USA (1010) *Using Evolutionary Couplings to Help Build and Refine Cryo-EM Models*
16:30–17:00	Pre Function	Coffee Available

WEDNESDAY, FEBRUARY 7 (continued)

17:00–19:15 **Mountain/Lake**

Cryo-EM of Large Macromolecular Assemblies

Kiyoshi Nagai, Medical Research Council, UK
Cryo-EM Snapshots of the Spliceosome Provide Insights into the Mechanism of Pre-mRNA Splicing.....7
Gira Bhabha, New York University School of Medicine, USA
Architectures of Lipid Transport Systems in Bacteria.....1
***Adam Frost**, University of California, San Francisco, USA
Function Follows Form: Cryo-EM Insights into Cell Biology.....N/A
Eva Nogales, University of California, Berkeley, USA
Challenges and Progress in the Study of Large Human Complexes in Gene Expression Regulation.....8

19:15–20:15 Granhall

Dinner

19:30–22:00 **Bay**

Poster Session 3

Poster sessions provide exciting opportunities for engagement between all levels of investigators. Abstracts beginning with the number 3 are featured during this poster session. Don't miss the plasma screen e-posters during this session!

20:00–21:00 Bay

Social Hour

THURSDAY, FEBRUARY 8

07:00–08:00 Granhall

Breakfast

08:00–11:15 **Mountain/Lake**

Cryo-EM of Membrane Proteins

***Georgios Skiniotis**, Stanford University, USA
Cryo-EM Visualization of G Protein-Coupled Receptors.....N/A
Judy Hirst, University of Cambridge, UK
CryoEM Structure of Mammalian Respiratory Complex I, an Asymmetric 1 MDa Energy-Converting Enzyme.....4

Coffee Break

Doreen Matthies, National Institutes of Health, USA
Single Particle Cryo-EM of a 200 kDa Magnesium Ion Channel.....6
Eric Gouaux, Oregon Health & Science University, USA
Structure and Mechanism of AMPA Receptor – Auxiliary Protein Complexes.....4
Kei Saotome, The Scripps Research Institute, USA (3012)
Short Talk: Structure of the Mechanically Activated Ion Channel Piezo1.....46
Luke H. Chao, Massachusetts General Hospital, USA (1017)
Short Talk: Single Particle Cryo-EM Studies of Mitochondrial Inner-Membrane Fusion.....19

On Own for Lunch

16:30–17:00 Pre Function

Coffee Available

*Session Chair; ( ) = abstract number

THURSDAY, FEBRUARY 8 (continued)

17:00–18:00	**Mountain/Lake**	**Recent Technological Advances in Cryo-EM** **Radostin Danev**, Max Planck Institute of Biochemistry, Germany *Single Particle Imaging with the Volta Phase Plate.....2* ***Bridget Carragher**, New York Structural Biology Center, USA *Sample Vitrification Issues and Some Ideas about How to Overcome Them.....1*
18:00–18:45	**Mountain/Lake**	**Closing Keynote Address** **Joachim Frank**, Columbia University, USA *Story in a Sample: Multiple Ribosome Structures in a Single Cryo-EM Experiment.....N/A*
18:45–19:00	**Mountain/Lake**	**Meeting Wrap-Up: Outcomes and Future Directions (Organizers)**
19:00–20:00	Granhall	Dinner
19:00–20:00	Granhall	Social Hour
20:00–23:00	Granhall	Entertainment

FRIDAY, FEBRUARY 9	Departure

Congratulations to Dr. Joachim Frank on his 2017 Nobel Prize and his contributions to the Cryo-EM field!

Thank you...

...to all our donors supporting this conference. Their generosity and dedication to the mission of collaborative science distinguish them as valuable members of the Keystone Symposia community.

This conference is sponsored by:

Pfizer Inc.

Additional support for this conference provided by:

Direct Electron, LP

Directors' Fund
(contributors noted by asterisks on blue pages)

Thermo Fisher Scientific

The following participants were awarded financial aid to attend this meeting. Be sure to stop by their poster sessions to see the abstracts that earned their merit-based support. Poster numbers are listed in parentheses, with the first digit indicating the corresponding poster session. Visit ***keystonesymposia.org/financialaid*** *for more information on Keystone Symposia's scholarship and travel award opportunities.*

The following scholarship recipients are funded by:

Keystone Symposia Future of Science Fund*

Halil Aydin, University of California, San Francisco, USA (1003)

Christina Heroven, MRC Laboratory of Molecular Biology, UK (1031)

Yuxi Liu, University of California, Los Angeles, USA (2015)

Vidya Mangala Prasad, University of Washington, USA (2023)

Conference Assistant

We would like to thank our conference assistant for contributing to this meeting. This individual assists the scientific organizers and Keystone Symposia's onsite staff throughout the meeting and compiles a meeting summary for submission to our government financial supporters.

Moran Shalev-Benami, Stanford University, USA (3014)

Meet the Scientific Organizers

Georgios Skiniotis, PhD is Professor of Molecular and Cellular Physiology and of Structural Biology at Stanford University's Bio-X institute and a structural biologist with expertise in electron cryo-microscopy (cryo-EM). Dr. Skiniotis has exploited the power of cryo-EM to study a wide range of important biological "machines" or macromolecular assemblies. His main interests are the mechanisms of transmembrane signal instigation with a particular focus on G protein-coupled receptors and cytokine receptors. The application of cryo-EM to such systems has also driven him to explore and refine approaches for resolving technically challenging problems.

Elizabeth Villa, PhD completed her PhD in Biophysics at the University of Illinois at Urbana-Champaign as a Fulbright Fellow. She was a Marie Curie Postdoctoral Fellow in the Max Planck Institute of Biochemistry in Munich before joining the Chemistry and Biochemistry Department faculty at the University of California, San Diego in July 2014. In 2016, she was awarded an NIH Director's New Innovator Award. She joined the Division of Biological Sciences faculty at UC San Diego as Assistant Professor in March 2017.

Her lab focuses on building tools for quantitative cell biology, using cryo-electron microscopy and tomography, cell biophysics, computational analysis and integrative modeling. Their biological focus is on the study of the nuclear periphery, as nuclear biology remains one of the most exciting challenges in the cell, and it is uncharted territory structurally. Projects include: the structural dynamics of the yeast nuclear pore complex, the mechanical communication between the cytoskeleton and the nucleus, and the molecular architecture of the genome and its association to the nuclear envelope. Dr. Villa also collaborates with different laboratories to open windows into various cellular events, including studying the inner life of bacteria and studying the effects of LRRK2 in Parkinson's disease, among others.

Andrew B. Ward, PhD is Assistant Professor in the Department of Integrative Structural and Computational Biology at The Scripps Research Institute in La Jolla, California. He and his laboratory work in a variety of areas including structural virology and membrane protein biology, employing many structural and biophysical techniques in order to provide details of these projects at a resolution ranging from atomic to ~50 Å. This broad range allows them to explore various facets of the biology of cellular and viral complexes associated with the membrane, and including influenza, HIV-1, hepatitis C, Ebola and others. They are interested in combining techniques such as x-ray crystallography, cryo-electron microscopy and biophysical assays with powerful computation in an integrative fashion. This philosophy enables the exploration of challenging structural problems where any single technique fails or is insufficient.

Dr. Ward's lab has developed a cryo-EM pipeline to provide rapid structural characterization of viral envelope glycoproteins in complex with monoclonal antibodies (mAb) at both low (15-20 Å) and high resolutions (4-10 Å), exploiting many recent advances in EM technology. Their custom pipeline and workflow enable them to address a variety of important biological and biomedical questions, including how viruses are recognized and neutralized by the human immune system, and how we can use the structural information to design novel or better vaccines. Recently, they solved the structure of the HIV-1 envelope glycoprotein trimer at 5.8 Å resolution, the type I fusion machinery responsible for host cell recognition and membrane fusion, and the target of the human immune response.

Thank you

To fulfill our nonprofit scientific mission, Keystone Symposia relies on significant financial commitment from government, foundation and industry donors as well as generous support from individuals. Gifts to the discretionary Directors' Fund ensure Keystone Symposia's ability to provide sufficient program support for meetings on a greater variety of scientific topics than would be possible if gifts were restricted to specific meetings in each case. All contributions are vital to the fulfillment of our mission — catalyzing collaborations to accelerate discoveries and helping to prepare and position the next generation of leading life scientists.

To all of our faithful supporters:
Thank you for your generous commitment to Keystone Symposia.

If your organization is interested in making a donation to Keystone Symposia, please contact our Development Office at ***development@keystonesymposia.org****. You can also visit* ***keystonesymposia.org/corporategiving*** *for more information.*

If you are interested in making a personal donation to the Future of Science Fund, please visit our secure individual donor platform at ***keystonesymposia.org/FSF****.*

To explore personal bequests and other planned gift vehicles, please contact our Development Office.

Donor Acknowledgement

Thank you...

Champions

Top-tier donors making an ongoing, annual commitment of $100,000+. Their public championing of Keystone Symposia's cause provides inspirational leadership commitment to our shared scientific mission of catalyzing collaborations, accelerating discoveries, and preparing and positioning the next generation of leading life scientists.

Bayer AG*
Cancer Research UK*
Croucher Foundation
Bill & Melinda Gates Foundation
Novo Nordisk A/S*
Pfizer Inc.*
Roche*
Science for Life Laboratory
The University of Hong Kong
Volkswagen Foundation
Knut and Alice Wallenberg Foundation

Sustaining Benefactors

Donors making a three-year or ongoing commitment of at least $50,000 per year. Their generous support is crucial to Keystone Symposia's ability to plan future scientific conferences focused on emerging topics and excellence in science.

AstraZeneca*
BioLegend, Inc.*
*Cell Research**
Incyte Corporation*
Takeda Pharmaceutical Company Limited*

Benefactors

Donors of $50,000 or above. We are very grateful for this extraordinary commitment to our mission to connect the scientific community and accelerate discoveries that benefit the world community. Special thanks to these organizations for consistent, annual Benefactor-level support.

Amgen Inc.*

Sustaining Sponsors

Donors making a three-year commitment of $25,000-$49,999 per year. Their generous support is crucial to Keystone Symposia's ability to plan future scientific conferences focused on emerging topics and excellence in science.

Astellas Pharma Inc.*
Bioverativ Therapeutics*
Editas Medicine*
EMD Serono Research and Development Institute, Inc.*
Gilead Sciences, Inc.*
Juno Therapeutics*
MedImmune*
Moderna Therapeutics*
MorphoSys AG*
Nestlé Institute of Health Sciences*
Novartis Institutes for BioMedical Research*
OncoMed Pharmaceuticals, Inc.*
Sangamo Therapeutics, Inc.*
Sanofi US*
Shire*
Surface Oncology*
TESARO, Inc.*
Thermo Fisher Scientific Inc.*

*Directors' Fund contributors; Information on this page is current as of January 15, 2018.

Donor Acknowledgement

Sponsors

Donors contributing $25,000-$49,999. These generous gifts allow us to convene meetings in a wide variety of important areas, many of which are in the early stages of research. Special thanks to these organizations for consistent, annual Sponsor-level support.

AbbVie Inc.*
Akebia Therapeutics*
Allen Institute for Brain Science
Biogen*
Bristol-Myers Squibb Company*
California Institute for Regenerative Medicine (CIRM)
ImmunoGen, Inc.*
Ionis Pharmaceuticals, Inc.*
Merck & Co., Inc.
Science Foundation Ireland
Theravance Biopharma*
Vertex Pharmaceuticals Incorporated*

Partners ($10,000–$24,999)

Baker Ruskinn
Diabetic Complications Consortium
Don Whitley Scientific Limited*
eLife*
HypOxygen*
Infinity Pharmaceuticals, Inc.*
Mercodia, Inc.*
*Oncotarget**
The Kavli Foundation
The Paul G. Allen Frontiers Group*
Thermo Fisher Scientific*
University of Notre Dame
Xcell Biosciences*

Patrons ($5,000–$9,999)

Abcam plc*
Adipogen International*
Agilent*
Agilent Technologies, Inc.
American Association for Cancer Research (AACR)*
American Society of Gene & Cell Therapy (ASGCT)*
Anaerobe Systems
Arraystar Inc.*
Avanti Polar Lipids, Inc.*
Bethyl Laboratories, Inc.*
Bio-Techne*
Charles River*
Chroma Technology Corporation*
Dharmacon, a Horizon Discovery Group Company*
Diversigen*
EpiCypher
Eurofins Pharma Discovery Services*
Immudex USA, LLC*
InSphero Inc.*
Integrated DNA Technologies*
Kenneth Rainin Foundation
Microbiome Insights, Inc.*
PeproTech, Inc.*
Synthego*
The Antibody Society
The Jackson Laboratory*
Zymo Research Corporation*

Donors ($2,500–$4,999)

ACS Infectious Diseases
ALPCO*
Cayman Chemical Company*
Cytokinetics, Inc.*
Fáilte Ireland
Laboratory Infrastructure Solutions
Novus Biologicals*
Promega Corporation*
Research Diets, Inc.*
Zymeworks

Contributors (up to $2,499)

BioElectron Technology Corporation
Direct Electron, LP
Forge Therapeutics, Inc.
Journal of General Physiology
New England Biolabs, Inc.
Opsona Therapeutics Ltd*
Spero Therapeutics
STEMCELL Technologies, Inc.*
The RNA Society

The generous support of our Partners, Patrons, Donors and Contributors makes possible the outstanding scientific quality of our meetings, and unsurpassed opportunities for interaction among attending scientists.

*Directors' Fund contributors; Information on this page is current as of January 15, 2018.

Donor Acknowledgement

Thank you…

Support for Keystone Symposia Diversity in Life Science Programs

We are grateful for this valuable support directed at increasing the participation of underrepresented scientists among meeting leaders and attendees, thereby enhancing diversity in the life science research community. Read more about our Diversity in Life Science programs by visiting: ***www.keystonesymposia.org/diversity****.*

Biogen
Burroughs Wellcome Fund
Endocrine Society
Regeneron Pharmaceuticals, Inc.

US Federal Grant Funding

Keystone Symposia truly appreciates the support received from various institutes of the National Institutes of Health. This support primarily funds scholarships for graduate students and postdoctoral fellows to attend our conferences. US federal grant support for the 2017–2018 meeting series was generously provided by:

Eunice Kennedy Shriver National Institute of Child Health and Human Development (NICHD)
National Center for Complementary and Integrative Health (NCCIH)
National Heart, Lung and Blood Institute (NHLBI)
National Institute on Aging (NIA)
National Institute of Allergy and Infectious Diseases (NIAID)
National Institute of Diabetes and Digestive and Kidney Diseases (NIDDK)
National Science Foundation (NSF)

Speaker Gift-In-Kind Donors

The following companies with speakers on one or more 2017–2018 Keystone Symposia meeting programs have generously agreed to forego reimbursements for speaker travel and lodging expenses.

Achaogen
Agios Pharmaceuticals
AstraZeneca
Circuit Therapeutics
Editas Medicine
Eli Lilly and Company
Entasis Therapeutics
Genentech, Inc.
Genmab A/S
GlaxoSmithKline Vaccines
Heptares Therapeutics Ltd
ImmunoGen, Inc.
Ionis Pharmaceuticals, Inc.
Janssen R&D: Pharmaceutical Companies of Johnson & Johnson
Lilly USA, LLC
Merck & Co., Inc.
Novartis Institutes for BioMedical Research
Pfizer Inc.
Pfizer Research Technology Center
RaNA Therapeutics Inc.
Regeneron Pharmaceuticals, Inc.
Sangamo Therapeutics, Inc.
Sanofi US
Spero Therapeutics
Thermo Fisher Scientific Inc.

Information on this page is current as of January 15, 2018.

In-Kind Media Partners

We are grateful to the following publishers and other organizations who have provided Keystone Symposia with advertising platforms to spread the word to those who can benefit from our meetings. If your organization is interested in an in-kind marketing/advertising partnership with Keystone Symposia, please contact Yvonne Psaila, Director of Marketing and Communications, at ***yvonnep@keystonesymposia.org*** *or* ***1.970.262.2676****.*

Platinum ($50,000+)

American Association for the Advancement of Science (AAAS)
BioMed Central Ltd
Cell Press
EMBO Press
The Scientist

Gold ($25,000–$49,999)

Nature Research

Silver ($10,000–$24,999)

Mary Ann Liebert, Inc. publishers
PLOS
The Rockefeller University Press – *The Journal of Cell Biology*
The Rockefeller University Press – *The Journal of Experimental Medicine*

Bronze ($2,500–$9,999)

Journal of Genomics, Proteomics and Bioinformatics
Journal of Pharmacological Research
Molecular Metabolism (published by Elsevier)
The Company of Biologists
The Journal of Clinical Investigation
The Journal of Rheumatology

Contributing (up to $2,500)

The Beatson Institute for Cancer Research

Information on this page is current as of January 15, 2018.

Keystone Symposia Future of Science Fund

These generous alumni of previous meetings and others with a passion for ensuring a future of scientific discovery that benefits humankind have made gifts during the last 12 months to support the Keystone Symposia Future of Science Fund. Through their generosity, we are able to provide scholarships and travel awards to the next generation of biomedical and life scientists, whose education and careers are enhanced by the opportunity to attend meetings and interact with the world's leading senior scientists.

Keystone Champions ($1,000–$4,999)

Hans-Rudolf Berthoud
Weizhi Ji
Sungwuk Kim
Walter Moos and Susan Miller
Craig Muir
Marco and Sofia
Kevan Shokat
Robert Tennant
Ian A. Wilson
Tachi and Leslie Yamada
Bei Zhang and Mark Erion
Anonymous (1)

Keystone Benefactors ($500–$999)

E. Dale Abel
Clement Adebamowo
Alberto Bosque*
John Cantwell with matching gift from Novartis
Sidney H. Cantwell Jr.
Peter Harris
Anke Huckriede
Shohei Koide
Antonio Lanzavecchia
Virginia Lee
Federica Sallusto

**Donors contributing on a recurring monthly basis*

View the most current listing of donors online at keystonesymposia.org/FSF; Information on this page is current as of January 15, 2018.

Keystone Symposia Future of Science Fund

Keystone Patrons ($100–$499)

Stephen Anderson
Alan Attie
Amos Baruch
Ann Belien
Bonnie B. Blomberg
Linda B. Bloom
Henrique Borges da Silva
Martha Carlin
Arup Chakraborty
Ajay Chawla
Rhea Coler
Peter Crawford
Pamela S. Daugherty*
Irina Dulubova
Joel Ernst
Dmitry Gabrilovich
Lucio Gama
Nancy L. Haigwood
Masataka Harada
Elizabeth Henske
Henry Higgs
Linda Hrycaj
Vijay K. Kalra
Hyoung-Il Kim
Joseph Kim
John Kirby
Masayasu Kojima
Jan E. Leach
Eric Lecuyer
Dong-Sup Lee
Yongjun Lee
Mary Loeken
Alessandro Magli
Gerard McGeehan
Vito Palombella
Alexander Pertsemlidis
Stefan Pflanz
In honor of Daphne Land Rothermel
Manfred Ruediger
Toshiro Saito
David Shapiro
Noriaki Shimizu
Heidi Daetwyler Simpson
Leonidas Stamatatos
Greg Suh
James Thomas
Koji Toshinai
Moriya Tsuji
Jeffery Twiss
Matthew Vander Heiden
Trudi Veldman
Walter Vernon
Angela Vincent
Brian Volkman
Alice Cort and Bruce Walker
Shaohui Wang
Yingxiao Wang
Stephanie Watts
Scott Weaver
Sally E. Wenzel
Masafumi Yoshinaga
Anonymous (2)

What Is the Future of Science Fund?

The Future of Science Fund is a way, through outright or estate gifts, to provide educational program or scholarship support to help fulfill Keystone Symposia's mission of catalyzing collaborations to accelerate discovery and of preparing and positioning the next generation of leading life scientists. Gifts to the Future of Science Fund can be designated to support:

- Competitive-based scholarships for students and postdoctoral fellows
- General conference program support

For more information on the Future of Science Fund or to make a gift, please visit **keystonesymposia.org/FSF** or contact the Development team at **development@keystonesymposia.org**.

View the most current listing of donors online at keystonesymposia.org/FSF; Information on this page is current as of January 15, 2018.

Donor Acknowledgement

Thank you...

Keystone Symposia Future of Science Fund

Keystone Contributors ($10–$99)

Rama Akondy
Jeffery Alexander
Alex Almasan
Juliano Alves
Warrison Andrade
Anet J. N. Anelone
Moshe Arditi
Alex Barrett
Christian Behl
Kendra Bence
Jonathan Bernardini
Kathrin Bernt
Elisabeth Bik
Raymond Blind
Thomas Boesen
Maria Rosa Bono
Ezio Bonvini
Heather Bradshaw
Derek Brazil
Elena Brenna
Michael Briskin
Matthew Brook
Peter Bunyard
Rotonya Carr
Linda S. Cauley
Teja Celhar
Sikander Chattha
Edmond Chan
Sunghwa Choe
Anette Christ
Lan Chu
Giovanni Ciriello
Andres Contreras
Elise Courtois
Anne Curtis
Angelo D'Alessandro
Darlene Dartt
Sanjoy Kumar Das
Maisie Dawes
Andrew DiNardo
Rosalie Driessen
Enrique Espinosa
Brett Etchebarne
Ursula Fearon
Jonathan Fox
Anu Gangopadhyay
Mathieu Garand
Lee Ann Garrett-Sinha
Evgenia V. Gerasimovskaya
Louis Gonzalez
Einar Hallberg
Paul Harris
Stephen Harris
Tamia Harris
Michael Holtzman
Samad Ibitokou
Brian Iritani
Mahita Jarjapu
Brian Jensen
Andrew Jones
Su Myung Jung
Yoshinori Katakura
Kim Kemper
Benedikt Kessler
Natalie Kiesling
Thomas Kietzmann
Mi-Kyung Kim
Dana-Lynn Koomoa
Alexander Kros
Irina Kufareva
Min-Hao Kuo
Paul Kyambadde
Kijun Lee
Lilach Lerman
Wei Li
Chung-Chih Lin
Philip Liu
Robert Lowery
Peetra Magnusson
Nancy Mah
Richard Maser
Johnjoe McFadden
Agnes Michel
Nitish Mittal
Annie Moisan
Michael Moser
Christine Moussion
Ronald Mugume
Leonard Nettey
Hironari Nishizawa
Rudolf Oehler
Daisuke Oikawa
Bilal Omer
Luis Ortiz
Muhammad Osman
Mario Ostrowski
Olga Pakhomova
Vijay Pancholi
Lena Pernas
Christopher Proud
Puneet Puri
Sufi Raja
Elena Rampanelli
Kathrin Renner
Jane Richardson
Gabrielle Romain
Matthew Rondina
Christopher E. Rudd
John Rumsey
Hannele Ruohola-Baker
Ma Isabel Salazar
Victoria Sandler
Ben Scheres
Ian Scott
Jean Scholz
Klaus Seuwen
Maria Skamagki
Ioakim Spyridopoulos
Cynthia Stauffacher
Meredith Steeves
Gary Strichartz
Hyun-Woo Suh
Noriyuki Sumida
Eva Szabo
Whit Tingley
Julian Trachsel
Jean-Francois Trempe
Edwin Van Amersfoort
Keith Van Haren
Charlotte Vines
Carston Wagner
Jing Wang
Qi Wang
Jeremy Wellen
Ingrid E. Wertz
Gregory Wiessner
Lisa Wilsbacher
Jed Wiltzius
Zhenbiao Yang
Todd Yeates
John Yu
Katherine Yutzey
Zhengxiao Zhang
Peiyao Zhao
Zhendong Zhao
Lan Zhou
Berislav Zlokovic
Anonymous (4)

View the most current listing of donors online at keystonesymposia.org/FSF; Information on this page is current as of January 15, 2018.

Poster Session 1

Monday, February 5 | 19:30–22:00
Bay

Presenter	Poster #	Abstract Title
Aller, Stephen	1001 *e-poster*	Structure of the Nematode Endosymbiont Xenorhabdus nematophilus Toxin Complex A at 3.1 A resolution.....10
Axelrod, Jeremy	1002	A laser-based Zernike phase plate for cryo-em: recent progress and future prospects.....10
Aydin, Halil	1003	The structural basis of regulated sphingolipid synthesis.....11
Barandun, Jonas	1004	Structure of the earliest stable precursor of the ribosome – the small subunit processome.....11
Bergeron, Julien	1005 *e-poster*	Structural characterization of the R-body complex, a macromolecular piston-like nano-machine.....12
Berndsen, Zachary	1006	Heterogeneity and Dynamics in the HIV-1 Envelope Glycan Shield.....12
Bigalke, Janna	1007	Cryo EM structure of the activated receptor tyrosine kinase RET signalling complex reveals the importance of its cysteine rich domain.....13
Boehning, Jan	1008	Structural analysis of a Parkinson's Disease LRRK2 mutant via cellular cryo-electron tomography and sub-tomogram averaging.....13
Botte, Mathieu	1033	Structure Based Drug Discovery of Membrane Protein Targets for Medicine that Matters.....14
Brasch, Julia	1009	Assembly of Clustered Protocadherin neuronal recognition complexes studied by X-ray crystallography and cryo-electron Tomography.....14
Brock, Kelly	1010	Using evolutionary couplings to help build and refine cryoEM models.....14
Buffalo, Cosmo	1011	Role of the dileucine motif in Nef-induced trimerization of the Arf1:AP-1 clathrin adaptor complex 15
Burrell, Anika	1012	Human IMPDH1 retinal degeneration mutations promote polymer assembly.....15
Butterwick, Joel	1013	Cryo-EM Structure of Orco: The Molecular Basis for Insect Olfaction.....16
Byrd, R.	1014	Design and Construction of Membrane Surface Complexes for Magnetic Resonance and CryoEM Studies.....16
Cash, Jennifer	1015	Structural Snapshots of a Dynamic Process in the Regulation of P-Rex, a Metastatic Factor.....17
Castano, Daniel	1037	Dynamo: flexible software tools for cryoelectron tomography data......17
Cavadini, Simone	1034	The molecular basis for UV-damage recognition in chromatin.....18
Chang, Wen	1016	Structure analyses of vaccinia viral A26 protein that is critical for mature virus endocytosis.....18
Chao, Luke	1017	Single particle Cryo-EM studies of mitochondrial inner-membrane fusion.....19
Chen, James	1018 *e-poster*	Super-resolution Single-Particle Electron Microscopy.....19
Chmielewski, David	1035	CryoEM structure of Mayaro virus: a re-emerging New World alphavirus.....19

Listing is in alphabetical order by presenter's last name.
Poster abstracts are viewable on the mobile app or on your Keystone Symposia online account.

Poster Session 1

Monday, February 5 | 19:30–22:00
Bay

Presenter	Poster #	Abstract Title
Christie, Michelle	1019	Deciphering the structure of a novel class of cholesterol-dependent cytolysin like proteins.....20
Clark, Lindsay	1020	Cryo-EM studies of the human sterol transporter G5G8 in nanodiscs.....20
Croxford, Matthew	1021 *e-poster*	Silicon Nitride as a Platform for Cryo-Electron Tomography Based Structural Cell Biology 21
Davies, Simon	1023	Use of Amphipathic Polymers in Preserving Native Lipids for Membrane Proteins in Cryo-EM.....21
Ekiert, Damian	1025	Architectures of lipid transport systems for the bacterial outer membrane.....22
Freyberg, Zachary	1026 *e-poster*	Ribosome-Associated Vesicles: a dynamic vesicular endoplasmic reticulum in secretory cells 22
Fu, Ziao	1027	Lipid Bilayer Structure in Native Cell Membrane Nanoparticles of Multidrug Exporter AcrB.....23
Gan, Lu	1028	Exploration of eukaryotic nuclei at molecular resolution.....23
Greber, Basil	1029	The cryo-EM structure of human TFIIH.....24
Harris, Audray	1030	Structures of nanoparticle immunogens for vaccine development.....24
He, Jun	1036	Embracing cryoEM in drug discovery.....25
Heroven, Christina	1031	In-situ structural studies of RPTPmu by cryo-electron tomography of FIB milled cells.....25
Hutchings, Joshua	1032	Cryo-EM tomography and subtomogram averaging of COPII assembled on membranes.....26
Kollman, Justin	1038	Polymerization of CTP synthase homologs in different filament geometries generates regulatory diversity.....33
Luo, Yu-Chun	1039	Purification of vaccinia viral a26 protein for biochemical analysis and crystallization.....36

Listing is in alphabetical order by presenter's last name.
Poster abstracts are viewable on the mobile app or on your Keystone Symposia online account.

Poster Session 2

Tuesday, February 6 | 19:30–22:00
Bay

Presenter	Poster #	Abstract Title
Hill, Christopher	2033	Structural basis for protein translocation by the Vps4 AAA ATPase.....25
Huynh, Kevin	2001 *e-poster*	CryoEM Structure of the Human Sodium-Coupled Acid-Base Transporter NBCe1.....26
Ivanov, Ivaylo	2002	Electron microscopy and integrative modeling shed light on the mechanisms of transcription initiation.....27
Januliene, Dovile	2003	Structure of the human MHC-I peptide-loading complex.....27
Kalia, Raghav	2004	Structural Basis of Mitochondrial Receptor Binding and GTP-Driven Conformational Constriction by Dynamin-Related Protein 1.....28
Kang, Jin Young	2005	Molecular mechanism of RNA hairpin-stabilized pause and pause inhibition by elongation factors NusG and RfaH during bacterial transcription.....28
Kasinath, Vignesh	2006	Cryo-EM structure of the full human PRC2 and regulation by AEBP2 and JARID2.....29
Kefauver, Jennifer	2007	Architecture of Volume-Regulated Anion Channels.....29
Khanna, Kanika	2008 *e-poster*	Understanding morphological changes accompanying sporulation in B. subtilis using in situ cryo-electron tomography.....30
Khayat, Reza	2009	Mechanism of cellular recognition and entry by porcine circovirus 2.....30
Kim, Doo Nam	2010 *e-poster*	CryoFIT: Flexible Fitting of Biomolecule Structures from Cryo-EM Maps with User Friendly Molecular Dynamics Simulation.....30
Kim, Laura	2011	The Ins and Outs of Benchmarking your cryo-TEM.....32
Kim, Seung Joong	2034	Structural and Functional Anatomy of a Nuclear Pore Complex.....31
Kirchdoerfer, Robert	2012	Structural insights into the SARS Spike-ACE2 receptor complex.....32
Kizziah, James	2013	Staphylococcus aureus pathogenicity islands use competing scaffolding proteins for the molecular pirating of bacteriophage capsids.....32
Liang, Bo	2035	Structure and Regulation of Respiratory Syncytial Virus RNA Polymerase.....33
Liu, Yuxi	2015	Near-atomic resolution cryo-EM for a small protein on a symmetric imaging scaffold.....34
Lo, Yu-Hua (Daisy)	2016	Structural Analysis Reveals the Features of Ribosome Assembly Factor Nsa1/WDR74 Important for Localization and Interaction with the AAA-ATPase Rix7/NVL2.....34
Low, Harry	2017	Architecture of a bacterial type II secretion system.....35
Lucato, Christina	2018	An integrated structural characterization of P-Rex1 activity in cancer cell signaling.....35
Ludlam, William	2019	Structure of the mLST8-CCT Folding Complex.....35
Luo, Min	2020	Cryo-EM reveals directional R-loop formation and substrate handover mechanisms in Type I CRISPR-Cas system.....N/A
Ma, Jianpeng	2021	Application of Parallel Continuous Simulated Tempering to Cryo-EM structural refinement at near-atomic resolution.....36

Listing is in alphabetical order by presenter's last name.
Poster abstracts are viewable on the mobile app or on your Keystone Symposia online account.

Poster Session 2

Tuesday, February 6 | 19:30–22:00
Bay

Presenter	Poster #	Abstract Title
Ma, Jinbiao	2022	Cryo-EM study of RISC loading complex Dicer2-R2D2.....37
Mangala Prasad, Vidya	2023	Using cryo-electron tomography and sub-tomogram averaging to study the integrated virus fusion machinery in intact influenza virions.....37
Martin, Gregory	2024	Anti-diabetic drug binding site in KATP channels revealed by Cryo-EM.....38
Martinez-Ortiz, Wilnelly	2025 *e-poster*	Conserved structural motifs can be used to improve cryoEM modeling of voltage gated ion channels.....38
Mishanina, Tatiana	2026	Global conformational changes in RNA polymerase accompany nascent RNA hairpin-stabilized transcriptional pause.....39
Morris, Kyle	2027	HIV-1 Nef as a cargo-selective switch in hijacking AP-1 clathrin adaptor complexes.....39
Navarro, Paula	2028	Corpora amylacea are dense membranous labyrinths containing intact mitochondria and vesicles.....39
Nguyen, Thi Hoang Duong	2029	Cryo-EM structure of human telomerase and new insights into its assembly and function.....40
Pallesen, Jesper	2031	Conformational Plasticity in Viral Glycoproteins.....40
Phillips, Chris	2032	Mechanistic insights into PIKK family function: High resolution cryoEM studies of ATM, ATR and DNA-PK provide a structural basis to understanding the regulation of the DNA damage response.....41
Porebski, Przemyslaw	2038	Molstack - platform for comparative presentation of density maps and their interpretations.....41
Pryor, Jr., Edward	2039	An update on detergent usage in Cryo-EM structure determination of membrane proteins......42
Roll-Mecak, Antonina	2040	Mechanistic insights from the cryo-EM visualization of the microtubule severing enzyme katanin.....45
Sterling, Sarah	2036	Structural characterization of septin filaments and their membrane association.....49
Turner, Hannah	2037	Visualizing polyclonal HIV vaccine responses using single particle electron microscopy.....52

Listing is in alphabetical order by presenter's last name.
Poster abstracts are viewable on the mobile app or on your Keystone Symposia online account.

Poster Session 3

Wednesday, February 7 | 19:30–22:00
Bay

Presenter	Poster #	Abstract Title
Pillon, Monica	3001	Structure of the Essential Eukaryotic Grc3/Las1 RNA Processing Machinery.....41
Puchades, Cristina	3002	Hand-over-hand: How ATP drives protein translocation.....42
Qu, Qianhui	3003	Structure of a histone H3K4 methyltransferase COMPASS.....43
Razi, Aida	3005	Ribosome biogenesis: emerging view for a central role of assembly factors in the maturation of the ribosomal subunits functional core using Cryo-EM.....43
Rheinberger, Jan	3006	Structural and Functional Characterization of a Cyclic Nucleotide-gated Ion Channel.....44
Richardson, Jane	3007	CaBLAM – A New Tool Designed to Validate 2.5-4Å Models from CryoEM.....44
Rodriguez, Alejandro	3008	Inhibitory Role of Functionalized Fullerenes in HIV-1 Maturation.....44
Roh, Soung Hun	3009	The 3.5 Å CryoEM Structure of Nanodisc Reconstituted Yeast Vacuolar ATPase Vo Proton Channel.....45
Rudack, Till	3010	ModelMaker: An interactive integrative modeling tool to build complete macromolecular machines.....45
Santosh, Vishaka	3011	Understanding AAV DNA packaging.....46
Saotome, Kei	3012	Structure of the Mechanically Activated Ion Channel Piezo1.....46
Schroeder, Gunnar	3013	Fibril structure of amyloid-ß(1-42) by cryo-electron microscopy.....47
Shalev-Benami, Moran	3014	Cryo-EM structure of the synaptic wiring protein Teneurin.....47
Shintre, Chitra	3015 ***e-poster***	Structure of the polycistic kidney disease TRP channel PC2.....48
Singharoy, Abhishek	3016 ***e-poster***	Molecular Dynamics Flexible Fitting at sub-5 Å Resolution....48
Southworth, Daniel	3036	Visualizing the Substrate Translocation Mechanism of the Hsp104 AAA+ Disaggregase......48
Su, Min	3017	Go Beyond Gctf in CTF Determination for Tilt Application in Single-Particle Cryo-EM.....50
Su, Zhaoming	3018	Electron Cryo-Microscopy Reveals Essential Ebola Nucleoprotein Interactions in Nucleocapsid-like Helical Assembly.....49
Sun, Chang	3019 ***e-poster***	Cryo-EM structure of alternative complex III/ aa3 cytochrome oxidase supercomplex from Flavobacterium johnsoniae.....50
Sun, Stella	3020	Flagellum couples cell shape to motility in Trypanosoma brucei.....51
Tsai, Chi-Hang	3039	Differential Vaccinia Virus entry and innate immune signaling in macrophages.....51
Tsutakawa, Susan	3021	Small angle X-ray scattering for data assisted protein structure prediction in CASP12 with prospects to improve accuracy of predictions.....52
Ubarretxena, Iban	3032 ***e-poster***	Structures of key intermediates in the human mitochondrial chaperonin reaction cycle.....53
Wang, Jinliang	3023	A Point Mutation that Leads to a Human Neurodegenerative Disorder Destabilizes the Heat Shock Protein 60 Chaperonin.....53

Listing is in alphabetical order by presenter's last name.
Poster abstracts are viewable on the mobile app or on your Keystone Symposia online account.

Poster Session 3

Wednesday, February 7 | 19:30–22:00
Bay

Presenter	Poster #	Abstract Title
Wang, Yuhang	3024 *e-poster*	Multisegment Molecular Dynamics Flexible Fitting for cryo-EM......53
Watanabe, Reika	3025	The in situ structural analysis of a pathogenic mutant LRRK2 protein in Parkinson's Disease using cryo-electron tomography.....54
Williams, James	3026	Visualizing mechanisms of influenza virus neutralization by intact IgG and Fab fragments requires complementary electron microscopy approaches.....54
Williamson, James	3033	Assembly of the Bacterial Large Ribosomal Subunit.....55
Woldeyes, Rahel	3027	Using X-ray Free Electron lasers to visualize solvent in the M2 proton channel.....55
Wu, Chao	3037	CryoEM structure and biochemical analysis reveal assembly and regulation mechanisms of Ebola nucleocapsid in viral RNA synthesis.....56
Wysocki, Vicki	3034	Native MS in Structural Biology: Surface Collisions of Protein and Nucleoprotein Complexes.....56
Xu, Pengbiao	3038	Structural insights into the role of phosphorylated IRF3 in IFN-β transcription.....57
Yokom, Adam	3028	Conformational dynamics reveal translocation mechanism of Hsp104, a yeast AAA+ disaggregase.....57
Zhang, Cheng	3035	Ribosome labeling: a method for high-resolution structural analysis of small RNAs using cryoEM.....58
Zhao, Rui	3029 *e-poster*	CryoEM structure of the yeast spliceosomal P complex at 3.3 Å resolution, the last major piece of the missing puzzle in the splicing cycle.....58
Zhou, Keda	3030	Single particle cryoEM reveals the specific recognition of centromeric nucleosomes by CENP-N.....59
Zhou, Qiangjun	3031	The Primed SNARE-Complexin-Synaptotagmin Complex for Neurotransmitter Release.....59

Listing is in alphabetical order by presenter's last name.
Poster abstracts are viewable on the mobile app or on your Keystone Symposia online account.

Melanie A. Adams-Cioaba
TTP Labtech Inc.
Cambridge, MA, USA
melanie.adams@ttplabtech.com

Robin Aglietti
Early Discovery Biochemistry
Genentech, Inc.
South San Francisco, CA, USA
aglietti.robin@gene.com

Stephen G. Aller
Pharmacology & Toxicology
University of Alabama at Birmingham
Birmingham, AL, USA

Stephen P. Anderson
Center for Advanced Biotechnology and Medicine
Rutgers University
Princeton, NJ, USA
anderson@cabm.rutgers.edu

Joseph W. Arndt
Physical Biochemistry
Biogen Idec
Cambridge, MA, USA
joe.arndt@biogen.com

Martina Audagnotto
University of California, San Diego
San Diego, CA, USA
maudagnotto@ucsd.edu

Agustin Avila Sakar
Gatan
Pleasanton, CA, USA
asakar@gatan.com

Jeremy Axelrod
Physics
University of California, Berkeley
Berkeley, CA, USA
jaxelrod@berkeley.edu

Halil Aydin
Biochemistry and Biophysics
University of California, San Francisco
San Francisco, CA, USA
hibaydin@gmail.com

Hyunsu Bae
Physiology, College of Korean Medicine
Kyung Hee University
Seoul, South Korea
hbae@khu.ac.kr

Yonghong Bai
Molecular Engineering
Amgen Inc.
Cambridge, MA, USA
ybai@amgen.com

David Baker
Biochemistry
University of Washington
Seattle, WA, USA
dabaker@uw.edu

Jonas Barandun
Laboratory of Protein and Nucleic Acid Chemistry
Rockefeller University
New York, NY, USA
jbarandun@rockefeller.edu

Joseph D. Batchelor
In Vitro Biology
Sanofi
Waltham, MA, USA
joe.batchelor@gmail.com

Charles E. Bell
Biological Chemistry and Pharmacology
Ohio State University
Columbus, OH, USA
bell.489@osu.edu

Julien R. C. Bergeron
Molecular Biology and Biotechnology
University of Sheffield
Sheffield, UK
j.bergeron@sheffield.ac.uk

Zachary Berndsen
Integrative Structural and Computational Biology
The Scripps Research Institute
La Jolla, CA, USA
zberndse@scripps.edu

Gira Bhabha
Skirball Institute
New York University School of Medicine
New York, NY, USA
gira@uchicago.edu

Janna M. Bigalke
Structure and Biophysics, Discovery Sciences
AstraZeneca
Gothenburg, Sweden
janna.bigalke@astrazeneca.com

Jan Boehning
Molecular Biology
University of California, San Diego
La Jolla, CA, USA
bojan@ucsd.edu

Thomas Boesen
Interdisciplinary Nanoscience Center
Aarhus University
Aarhus, Denmark
thb@mbg.au.dk

Kenneth Borrelli
Life Science Software
Schrodinger Inc.
New York, NY, USA
kenneth.borrelli@schrodinger.com

Mathieu Botte
leadXpro AG
Villigen, Switzerland
mathieu.botte@leadxpro.com

Julia Brasch
Biochemistry and Molecular Biophysics
Columbia University
New York, NY, USA
jb2754@columbia.edu

Jacob Brink
Sales
JEOL USA, Inc.
Peabody, MA, USA
jbrink@jeol.com

Kelly Brock
Cell Biology and Systems Biology
Harvard Medical School
Boston, MA, USA
kpgbrock@gmail.com

Christopher Browning
Vertex Pharmaceuticals
Abingdon, UK
christopher_b_browning@vrtx.com

Cosmo Zephyrus Buffalo
Cellular and Molecular Biology
University of California, Berkeley
Berkeley, CA, USA
buffalo@berkeley.edu

Anika L. Burrell
Biochemistry
University of Washington
Seattle, WA, USA
anikab@uw.edu

Joel Butterwick
Rockefeller University
New York, NY, USA
jbutterwic@rockefeller.edu

R. Andrew Byrd
Structural Biophysics Laboratory
NCI, National Institutes of Health
Frederick, MD, USA
byrdra@mail.nih.gov

Elliot Campbell
Macrotope, Inc.
Somerset, NJ, USA
elliot@macrotope.com

Meeting Participant List (current as of January 15, 2018)

Bridget Carragher
New York Structural Biology Center
New York, NY, USA
bcarr@nysbc.org

Stephen David Carter
Biology
California Institute of Technology
Pasadena, CA, USA
scarter@caltech.edu

Jennifer Cash
Life Sciences Institute
University of Michigan
Ann Arbor, MI, USA
cashjn@umich.edu

Daniel Castano
University of Basel
Basel, Switzerland
daniel.castano@unibas.ch

Simone Cavadini
Friedrich Miescher Institute
Basel, Switzerland
simone.cavadini@fmi.ch

Frederic Cazals
Algorithms-Biology-Structure
Institut National de la Recherche Agronomique
Sophia-Antipolis, France
frederic.cazals@inria.fr

Tom Ceska
Director of Crystallography
UCB Celltech
Slough, UK
Tom.Ceska@ucb.com

Wen Chang
Institute of Molecular Biology
Academia Sinica
Taipei, Taiwan
mbwen@ccvax.sinica.edu.tw

Luke H. Chao
Molecular Biology
Massachusetts General Hospital
Boston, MA, USA
chao@molbio.mgh.harvard.edu

Ines Chen
Nature Structural and Molecular Biology
Nature Publishing Group
New York, NY, USA
i.chen@us.nature.com

James Chen
Oregon Health & Science University
Portland, OR, USA

Xi Chen
Chemistry
Northwest University
Xi'an, China
sealasting@hotmail.com

David Chmielewski
Biophysics
Stanford University
Palo Alto, CA, USA
David.Chmielewski@stanford.edu

Lindsay Clark
Biophysics
University of Texas Southwestern Medical Center
Dallas, TX, USA
lindsay.clark@utsouthwestern.edu

Sergio J. Couoh-Cardel
Molecular and Cellular Physiology
Stanford University
Menlo Park, CA, USA
scouoh@stanford.edu

Patrick Cramer
Molecular Biology
Max Planck Institute for Biophysical Chemistry
Goettingen, Germany
patrick.cramer@mpibpc.mpg.de

Matthew Croxford
Biological Sciences
University of California, San Diego
San Diego, CA, USA
mcroxfor@ucsd.edu

Radostin Danev
Molecular Structural Biology
Max Planck Institute of Biochemistry
Martinsried, Germany
danev@biochem.mpg.de

Simon Davies
Faculty of Biological Sciences
University of Leeds
Leeds, West Yorkshire, UK
bsspd@leeds.ac.uk

Cony D'Cruz
Schrodinger, Inc.
Portland, OR, USA
cynthia.pariente@schrodinger.com

Natalie de Souza
Nature Methods
New York, NY, USA
n.desouza@us.nature.com

Hasan Demirci
Biosciences Division/PULSE Institute/ SLAC
Stanford University
Menlo Park, CA, USA
Hasan_Demirci@Stanford.edu

Joao M. Dias
SMS
Pfizer Inc.
Groton, CT, USA
jmdias@pfizer.com

Xinlin Du
Structural Biology
Peloton Therapeutics
Dallas, TX, USA
xinlin.du@pelotontx.com

Damian C. Ekiert
University of Chicago
New York, NY, USA
dcekiert@uchicago.edu

Mark H. Ellisman
National Center for Microscopy and Imaging Research
University of California, San Diego
La Jolla, CA, USA
mark@ncmir.ucsd.edu

Joachim Frank
Biochemistry and Molecular Biophysics
Columbia University
New York, NY, USA
jf2192@columbia.edu

Matthew C. Franklin
Structural Biology
Regeneron Pharmaceuticals
Tarrytown, NY, USA

N. Freeman
National Institutes of Health
Bethesda, MD, USA

Zachary Freyberg
Psychiatry
University of Pittsburgh
Pittsburgh, PA, USA
freyberg@pitt.edu

Adam Frost
Biochemistry and Biophysics
University of California, San Francisco
San Francisco, CA, USA
adam.frost@ucsf.edu

Ziao Fu
Integrated Program
Columbia University
New York, NY, USA
zf2147@cumc.columbia.edu

Michael A. Funk
AAAS Science
Washington, DC, USA
mfunk@aaas.org

Lu Gan
Biological Sciences
National University of Singapore
Singapore, Singapore
lu@anaphase.org

Stephanie Gates
University of California, Berkeley
Berkeley, CA, USA
sngates@berkeley.edu

Yacob Gomez Llorente
Structural Chemistry
Merck
Kenilworth, NJ, USA
yacob.gomez.llorente@merck.com

Eric Gouaux
Vollum Institute
Oregon Health & Science University
Portland, OR, USA
gouauxe@ohsu.edu

Basil J. Greber
QB3
University of California, Berkeley
Berkeley, CA, USA
basilgreber@berkeley.edu

Kay Grünewald
Nuffield Department of Medicine
University of Oxford
Headington, Oxford, UK
kay@strubi.ox.ac.uk

Keelan Z. Guiley
Chemistry
University of California, Santa Cruz
Los Gatos, CA, USA
kguiley@ucsc.edu

Fei Guo
Molecular Cellular Biology
University of California, Davis
Davis, CA, USA
feiguo@ucdavis.edu

Susan Hafenstein
Medicine
Pennsylvania State University
Hershey, PA, USA
shafenstein@psu.edu

Seungil Han
Structural and Molecular Sciences
Pfizer Inc.
Groton, CT, USA
seungil.han@pfizer.com

Eric Hanssen
Bio21 Institute
University of Melbourne
Melbourne, VIC, Australia
ehanssen@unimelb.edu.au

Audray K. Harris
NIAID, National Institutes of Health
Bethesda, MD, USA
harrisau@mail.nih.gov

Rachel C. Hart
Pathology, Microbiology, and Immunology
Vanderbilt University Medical Center
Nashville, TN, USA
rachel.c.hart.1@vanderbilt.edu

Jun He
UCB
Slough, Berkshire, UK
Jun.He@ucb.com

Gregor Heiss
Thermo Fisher Scientific
Wien, Austria
gregor.heiss@fei.com

Christina Heroven
MRC Laboratory of Molecular Biology
Cambridge, UK
cheroven@mrc-lmb.cam.ac.uk

Christopher P. Hill
Biochemistry
University of Utah
Salt Lake City, UT, USA
chris@biochem.utah.edu

Joel A. Hirsch
Biochemistry
Tel Aviv University
Tel Aviv, Israel
jhirsch@post.tau.ac.il

Judy Hirst
MRC Mitochondrial Biology Unit
University of Cambridge
Cambridge, UK
jh@mrc-mbu.cam.ac.uk

Heather Holdaway
Thermo Fisher Scientific
New Castle, PA, USA
heather.holdaway@thermofisher.com

David J. Hosfield
Ben May Institute for Cancer Research
University of Chicago
Chicago, IL, USA
hosfield@uchicago.edu

Joshua Hutchings
Biological Sciences
Birkbeck College
London, UK
joshua.hutchings.15@ucl.ac.uk

Kevin Huynh
Microbiology, Immunology & Molecular Genetics
University of California, Los Angeles
Los Angeles, CA, USA
kwhuynh@ucla.edu

Artur Indzhykulian
Harvard University
Boston, MA, USA
Artur_Indzhykulian@hms.harvard.edu

Ivaylo Ivanov
Chemistry
Georgia State University
Atlanta, GA, USA
iivanov@gsu.edu

Dovile Januliene
Structural Biology
Max Planck Institute of Biophysics
Frankfurt, Germany
Dovile.Januliene@biophys.mpg.de

Christine Jao
Structural Biology
Genentech, Inc.
South San Francisco, CA, USA
jao.christine@gene.com

Grant J. Jensen
Biology
California Institute of Technology
Pasadena, CA, USA
jensen@caltech.edu

M. Gordon Joyce
U.S. Military HIV Research Program
Bethesda, MD, USA
gjoyce@hivresearch.org

Raghav Kalia
Biochemistry
University of Utah
Salt Lake City, UT, USA

Jin Young Kang
Rockefeller University
New York, NY, USA
jkang@rockefeller.edu

Meeting Participant List (current as of January 15, 2018)

Vignesh Kasinath
QB3 Institute
University of California, Berkeley
Berkeley, CA, USA
vigneshk@berkeley.edu

Jennifer Kefauver
Integrative Structural and Computational Biology
The Scripps Research Institute
La Jolla, CA, USA
kefauver@scripps.edu

Kanika Khanna
Division of Biological Sciences
University of California, San Diego
San Diego, CA, USA
kkhanna@ucsd.edu

Reza Khayat
Chemistry and Biochemistry
City College of New York
New York, NY, USA
rkhayat@ccny.cuny.edu

Doo Nam Kim
Biophysics
Los Alamos National Laboratory
Los Alamos, NM, USA
doonam@lanl.gov

Laura Y. Kim
Simons Electron Microscopy Center
New York Structural Biology Center
New York, NY, USA
lkim@nysbc.org

Russell King
TTP Labtech Ltd.
Melbourn, Herts, UK
russell.king@ttplabtech.com

Robert N. Kirchdoerfer
Immunology and Microbial Sciences
The Scripps Research Institute
La Jolla, CA, USA
rkirchdo@scripps.edu

James L. Kizziah
Microbiology
University of Alabama at Birmingham
Birmingham, AL, USA
kizziah4@uab.edu

Justin M. Kollman
Biochemistry
University of Washington
Seattle, WA, USA
jkoll@uw.edu

Floyd M. Kregenow
Biochemistry
University of Washington
Lynnwood, WA, USA
fkregenow@gmail.com

Ira Kurtz
Medicine, Nephrology
University of California, Los Angeles
Los Angels, CA, USA
Ikurtz@mednet.ucla.edu

Vinson Lam
Division of Biology
University of California, San Diego
San Diego, CA, USA
vclam@ucsd.edu

Gabriel C. Lander
Integrative Structural and Computational Biology
The Scripps Research Institute
La Jolla, CA, USA
glander@scripps.edu

Matthias E. Lauer
Chemical Biology
Roche Innovation Center Basel
Basel, Switzerland
Matthias.Lauer@roche.com

Bryden Le Bailly
Nature
London, UK
bryden.lebailly@nature.com

Hyun Lee
University of Illinois at Chicago
Chicago, IL, USA
danielhl@uic.edu

Jeremy Leitz
Molecular and Cellular Physiology
Stanford University
Stanford, CA, USA
jleitz@stanford.edu

Bo Liang
Biochemistry
Emory University School of Medicine
Atlanta, GA, USA
bo.liang@emory.edu

Huiyang Winnie Liang
Chemistry
Stanford University
Palo Alto, CA, USA
wliang6@stanford.edu

Maofu Liao
Cell Biology
Harvard Medical School
Boston, MA, USA
maofu_liao@hms.harvard.edu

Scott E. Lindner
Biochemistry and Molecular Biology
Pennsylvania State University
University Park, PA, USA
scott.lindner@psu.edu

Jinsong Liu
Guangzhou Institutes of Biomedicine and Health
Chinese Academy of Sciences
Guangzhou, China
liu_jinsong@gibh.ac.cn

Jun Liu
Microbial Pathogenesis
Yale School of Medicine
West Haven, CT, USA
jun.liu.jl2996@yale.edu

Yuxi Liu
Chemistry and Biochemistry
University of California, Los Angeles
Los Angeles, CA, USA
yuxiliu@chem.ucla.edu

Yu-Hua (Daisy) Lo
NIEHS, National Institutes of Health
Durham, NC, USA
yu-hua.lo@nih.gov

Harry Low
Life Sciences
Imperial College
London, UK
h.low@imperial.ac.uk

Christina M. Lucato
Monash University
Glen Huntly, Victoria, Australia
christina.lucato@monash.edu

William G. Ludlam
Chemistry/Biochemistry
Brigham Young University
Provo, UT, USA
w.grant.ludlam@gmail.com

Min Luo
Cell Biology
Harvard Medical School
Boston, MA, USA
min_luo@hms.harvard.edu

Yu-Chun Luo
Institute of Molecular Biology
Academia Sinica
Taipei, Taiwan
luoyuchun@gate.sinica.edu.tw

Dmitry Lyumkis
Genetics
The Salk Institute for Biological Studies
La Jolla, CA, USA
dlyumkis@salk.edu

Jianpeng Ma
Biochemistry and Molecular Biology
Baylor College of Medicine
Houston, TX, USA
jpma@bcm.tmc.edu

Jinbiao Ma
Biochemistry
Fudan University
Shanghai, China
majb@fudan.edu.cn

Yingli Ma
Chemistry
Amgen Inc.
Shanghai, China
yinglim@amgen.com

Vidya Mangala Prasad
Medicinal Chemistry
University of Washington
Seattle, WA, USA
vmprasad@uw.edu

Gregory Martin
Biochemistry and Molecular Biology
Oregon Health & Science University
Portland, OR, USA
martingr@ohsu.edu

Wilnelly Martinez-Ortiz
Biochemistry and Molecular Pharmacology
New York University, School of Medicine
New York, NY, USA

Doreen Matthies
Laboratory of Cell Biology
National Institutes of Health
Bethesda, MD, USA
doreen.matthies@nih.gov

Trevor Allen McQueen
Chemistry
Stanford University
Palo Alto, CA, USA
trevor.allen.mcqueen@gmail.com

Xiaoshan Min
Molecular Engineering
Amgen Inc.
South San Francisco, CA, USA
xmin@amgen.com

Tatiana V. Mishanina
Biochemistry
University of Wisconsin
Madison, WI, USA
mishanina@wisc.edu

Jonathan M. Moore
Biology
Vertex Pharmaceuticals Incorporated
Boston, MA, USA
jonathan_moore@vrtx.com

Kyle L. Morris
QB3
University of California, Berkeley
San Francisco, CA, USA
kylelmorris@berkeley.edu

Tinoush Moulaei
MTS
USAMRIID
Frederick, MD, USA
tmoulaei@genevausa.org

Kiyoshi Nagai
Lab of Molecular Biology
Medical Research Council
Cambridge, UK
kn@mrc-lmb.cam.ac.uk

Eswar Narayanan
DuPont Pioneer
Johnston, IA, USA
narayanan.eswar@pioneer.com

Paula Perez Navarro
Center for Cellular Imaging and Nanoanalyitics
Biozentrum, University of Basel
Basel, Switzerland
pau.pereznavarro@unibas.ch

Thi Hoang Duong Nguyen
Molecular and Cell Biology
University of California, Berkeley
Berkeley, CA, USA
knguyen289@gmail.com

Daniela Nicastro
Cell Biology and Biophysics
University of Texas Southwestern Medical Center
Dallas, TX, USA
daniela.nicastro@utsouthwestern.edu

Ingrid Niesman
Biology
San Diego State University
San Diego, CA, USA

Eva Nogales
Molecular and Cell Biology
University of California, Berkeley
Berkeley, CA, USA
enogales@lbl.gov

Akihiro Nomura
Pharmaceutical Frontier Research Laboratries
JT Inc.
Yokohama, Kanagawa, Japan

Rie Omi
Medicinal Chemistry
ONO Pharmaceutical CO., LTD.
Los Angeles, CA, USA
oumi@ono.co.jp

Jesper Pallesen
The Scripps Research Institute
San Diego, CA, USA
pallesen@scripps.edu

Gabor Papai
Integrated Structural Biology
Institut de Génétique et de Biologie Moléculaire
et Cellulaire
Illkirch, France
papai@igbmc.fr

Daniel Perl
Pathology
Uniformed Services University of the Health Sciences
Bethesda, MD, USA
daniel.perl@usuhs.edu

John J. Peters
Molecular Cellular Physiology
Stanford University
Stanford, CA, USA
johnjp@stanford.edu

Chris Phillips
Discovery Sciences
AstraZeneca
Cambridge, UK
chris.phillips@astrazeneca.com

Edith Pierre-Jerome
Biology
Duke University
Durham, NC, USA
edith.pierre.jerome@duke.edu

Monica Pillon
NIEHS, National Institutes of Health
Durham, NC, USA
monica.pillon@nih.gov

Meeting Participant List (current as of January 15, 2018)

Przemyslaw Porebski
Molecular Physiology and Biological Physics
University of Virginia
Charlottesville, VA, USA
pjp2b@virginia.edu

Edward E. Pryor, Jr.
Anatrace
Maumee, OH, USA
edward_pryor@anatrace.com

Cristina Puchades
DiscoBio
The Scripps Research Institute
La Jolla, CA, USA
puchades@scripps.edu

Qianhui Qu
Molecular and Cellular Physiology
Stanford University
Stanford, CA, USA
qqh@stanford.edu

Aida Razi
Anatomy and Cell Biology
McGill University
Montreal, QC, Canada
aida.razi@mail.mcgill.ca

Jan Rheinberger
Anesthesiology
Weill Cornell Medicine
New York, NY, USA
jar2075@med.cornell.edu

Jane S. Richardson
Biochemistry
Duke University
Durham, NC, USA
jsr@kinemage.biochem.duke.edu

Alejandro Rodriguez
Chemistry
University of Texas at El Paso
El Paso, TX, USA
arodriguez109@miners.utep.edu

Soung Hun Roh
Bioengineering
Stanford University
Palo Alto, CA, USA
shroh@stanford.edu

Till Rudack
Biophysics
Ruhr University Bochum
Bochum, Germany
till.rudack@rub.de

Andrej Sali
Bioengineering and Therapeutic Sciences
University of California, San Francisco
San Francisco, CA, USA
sali@salilab.org

Vishaka Santosh
Physiology and Biophysics
Virginia Commonwealth University
Richmond, VA, USA
santoshv@vcu.edu

Kei Saotome
Integrative Computational and Structural Biology
The Scripps Research Institute
La Jolla, CA, USA
saotome@scripps.edu

Raymond Schrijver
Thermo Fisher Scientific
Eindhoven, Netherlands
Raymond.Schrijver@fei.com

Gunnar Schroeder
Structural Biochemistry
Forschungszentrum Julich
Juelich, Germany

Heidi Schubert
Biochemistry
University of Utah
Salt Lake City, UT, USA
heidi@biochem.utah.edu

Paulina Seweryn
Molecular Biology
Max Planck Institute for Biophysical Chemistry
Goettingen, Germany
paulina.seweryn@mpibpc.mpg.de

Moran Shalev-Benami
Molecular & Cellular Physiology
Stanford University
Stanford, CA, USA
moransb@stanford.edu

John Shelley
Schrodinger Inc.
Portland, OR, USA
John.Shelley@schrodinger.com

Mee Shelley
Drug Discovery Group
Schrodinger Inc.
Portland, OR, USA
Mee.Shelley@schrodinger.com

Chitra Shintre
Nuffield Department of Medicine
University of Oxford
Oxford, UK
chitra.shintre@sgc.ox.ac.uk

Kailene Simon
Rare Disease Research
Sanofi
Framingham, MA, USA
kailene.simon@sanofi.com

Digvijay Singh
Baltimore, MD, USA
digvijay.in1@gmail.com

Abhishek Singharoy
Biodesign Institute
Arizona State University
Tempe, AZ, USA
abhi@ks.uiuc.edu

Jitin Singla
Computational Biology and Bioinformatics
University of Southern California
Los Angeles, CA, USA
jsingla@usc.edu

Georgios Skiniotis
Molecular and Cellular Physiology
Stanford University
Stanford, CA, USA
yiorgo@stanford.edu

Daniel Southworth
Biochemistry and Biophysics
University of California, San Francisco
San Francisco, CA, USA
daniel.southworth@ucsf.edu

Michael Spilman
Direct Electron, LP
San Diego, CA, USA
mspilman@directelectron.com

Sarah M. Sterling
Molecular and Cell Biology
University of California, Berkeley
Berkeley, CA, USA
smsterling@berkeley.edu

Min Su
Life Sciences Institute
University of Michigan
Ann Arbor, MI, USA
minsu@umich.edu

Zhaoming Su
Bioengineer, SLAC
Stanford University
Stanford, CA, USA
suzm@stanford.edu

Chang Sun
Beckman Institute
University of Illinois
Urbana, IL, USA
sun.changing@gmail.com

Ming Sun
University of California, San Francisco
San Francisco, CA, USA
ming.sun@ucsf.edu

Stella Ying Sun
Bioengineering
Stanford University
Palo Alto, CA, USA
sying08@u.nus.edu

Chi-Hang Tsai
Institute of Molecular Biology
Academia Sinica
Taipei, Taiwan
jason0409@gate.sinica.edu.tw

Kuang-Lei Tsai
Biochemistry and Molecular Biology
University of Texas Health Science Center at Houston
Houston, TX, USA
kuang-lei.tsai@uth.tmc.edu

Yoshiro Tsuji
Chemical Research Laboratories
Kyowa Hakko Kirin Co., Ltd.
Shizuoka, Japan
yoshiro.tsuji@kyowa-kirin.co.jp

Susan E. Tsutakawa
Life Science Division
Lawrence Berkeley National Laboratory
Berkeley, CA, USA
setsutakawa@lbl.gov

Hannah L. Turner
Integrative Structural and Computational Biology
The Scripps Research Institute
San Diego, CA, USA
hannaht@scripps.edu

Iban Ubarretxena
Pharmacological Sciences
Icahn School of Medicine at Mount Sinai
New York, NY, USA
iban.ubarretxena@gmail.com

Carla Cristina Uranga
Research and Development
PaxVax
San Diego, CA, USA
curanga@paxvax.com

Elizabeth Villa
Chemistry and Biochemistry
University of California, San Diego
La Jolla, CA, USA
evilla@ucsd.edu

Clarissa Villinger
Central Facility of Electron Microscopy
Ulm University
Ulm, Germany

Laurent Michel Vuillard
Protein Chemistry
Institut de Recherches Servier
Criossy sur Seine, France
laurent-michel.vuillard@servier.com

Hongpeng Wang
Chemistry and Biochemistry
University of Maryland College Park
College Park, MD, USA
hongpeng@umd.edu

Jinliang Wang
Chemistry
University of Texas at El Paso
El Paso, TX, USA
jwang7@miners.utep.edu

Yuhang Wang
Biophysics
University of Illinois at Urbana-Champaign
Urbana, IL, USA
ywang148@illinois.edu

Andrew B. Ward
Integrative Structural and Computational Biology
The Scripps Research Institute
La Jolla, CA, USA
andrew@scripps.edu

Reika Watanabe
Molecular Biology
University of California, San Diego
La Jolla, CA, USA
recastillon@ucsd.edu

Ariel Hannah Wein
Chemistry
University of Southern California
Los Angeles, CA, USA
awein@usc.edu

Jui-Hung Weng
Pharmacology
University of California, San Diego
La Jolla, CA, USA
jhw043@ucsd.edu

Kristopher Ian White
Molecular and Cellular Physiology/HHMI
Stanford University
Stanford, CA, USA
kiwhite@stanford.edu

Christian Wiesmann
NIBR
Novartis
Basel, Switzerland
christian.wiesmann@novartis.com

Barry M. Willardson
Chemistry and Biochemistry
Brigham Young University
Provo, UT, USA
bmwillardson@chem.byu.edu

Christopher J. Williams
Biochemistry
Duke University
Durham, NC, USA
christopher.sci.williams@gmail.com

James Alan Williams
Medicinal Chemistry
University of Washington
Seattle, WA, USA
jamesw24@uw.edu

James R. Williamson
Integrative Structural and Computational Biology
The Scripps Research Institute
La Jolla, CA, USA
jrwill@scripps.edu

Steven C. Wilson
MCP
Stanford University
Stanford, CA, USA

Sharon Grayer Wolf
Chemical Research Support
Weizmann Institute of Science
Rehovot, Israel
sharon.wolf@weizmann.ac.il

Chao Wu
Washington University
St. Louis, MO, USA
chaowu@wustl.edu

Meeting Participant List (current as of January 15, 2018)

Mary Ann Wu
National Institute of General Medical Sciences
National Institutes of Health
Bethesda, MD, USA
maryann.wu@nih.gov

Vicki H. Wysocki
Chemistry and Biochemistry
Ohio State University
Columbus, OH, USA
wysocki.11@osu.edu

Pengbiao Xu
Biochemistry and Biophysics
Texas A&M University
College Station, TX, USA
xpb20081@tamu.edu

Su Yang
The Scripps Research Institute
La Jolla, CA, USA
syang@scripps.edu

Mark Yeager
Molecular Physiology and Biological Physics
University of Virginia School of Medicine
Charlottesville, VA, USA
yeager@virginia.edu

Todd Yeates
Chemistry and Biochemistry
University of California, Los Angeles
Los Angeles, CA, USA
yeates@mbi.ucla.edu

Adam Lee Yokom
Molecular and Cell Biology
University of California, Berkeley
Berkeley, CA, USA
alyokom@berkeley.edu

Lindsey Young
Molecular & Cellular Biology
University of California, Berkeley
Berkeley, CA, USA
lindsey.young@berkeley.edu

Cheng Zhang
The Salk Institute for Biological Studies
San Diego, CA, USA
chzhang@salk.edu

Kam Zhang
Center for Life Science Technologies
RIKEN
Yokohama, Kanagawa, Japan
kamzhang@riken.jp

Rui Zhao
Biochemistry and Molecular Genetics
University of Colorado
Denver, CO, USA
rui.zhao@ucdenver.edu

Keda Zhou
Chemistry and Biochemistry
University of Colorado Boulder
Boulder, CO, USA
keda.zhou@colorado.edu

Keystone Symposia Staff

Jane Peterson, Chief Executive Officer
Pamela Daugherty, Chief Financial Officer
Thale Jarvis, Chief Scientific Officer
Rick Sherman, Vice President, Development
Jeannie Dalrymple, Senior Director, Global Development
Nick Dua, Senior Director, Communications
Linda Hrycaj, Senior Director, Strategic Planning
Kevin Aoki, Director, Information Technology
Sarah Lavicka, Director, Development
Allison Ogdon, Director, Program Development & Implementation
Yvonne Psaila, Director, Marketing & Communications
Irelene Ricks, Director, Diversity in Life Science Programs
Mary Jo Roal, Director, Human Resources
Heidi Daetwyler Simpson, Director, Meeting Management

David Adamson, On-Site Staff Representative
Cathy Banks, Web Developer/Systems Analyst
Patricia Beard, Attendee Services Specialist
Carol Bosserman, On-Site Staff Representative
Mary Brown, On-Site Staff Representative
Caroline Brendel, Executive Administrative Assistant
Doug Castle, Web Developer/Systems Analyst
Chris Dean, IT Support Specialist
Heather Ford, Manager, Attendee Services
Marcus Fresia, Associate Conference Coordinator
Lindsey Heilmann, Conference Coordinator
Jenny Hindorff, Program Development Coordinator
Kristina Horton Richardson, SAB Coordinator/Administrative Assistant
Jeff Lehman, Scholarship Coordinator
Kellie McConnell, Program Implementation Coordinator
Alexa Morrison, Program Implementation Associate
Tanya Muller, Lead Software Developer
Pauli Novak, Marketing & Meeting Book Production Specialist
Annie Page, On-Site Staff Representative
Kelly Sauerwein, Grant Coordinator
Barbara Schmid-Miller, Marketing & Social Media Specialist
Julie Schmidt, On-Site Staff Representative
Dawn Shafer, Accounting Clerk
Kiki Sperr, Development Coordinator
Bridgett Thompson, Logistics Specialist
Rebecca Wilkerson, Finance Assistant

To reach one of these staff members, please call 1-800-253-0685 or 1-970-262-1230.

Keystone Symposia Boards and Consultants

Board of Directors *Members of the Board of Directors are also members of the Scientific Advisory Board.*

Scientific Advisory Board

Keystone Symposia Boards and Consultants

Scientific Advisory Board (continued)

The Board of Directors and Scientific Advisory Board listings reflect members at the time of printing.

Scientific Advisory Board (continued)

Please visit the following link to view the Keystone Symposia Programming Consultants who contributed to the current season's conference program development: **www.keystonesymposia.org/consultants**

The Board of Directors and Scientific Advisory Board listings reflect members at the time of printing.

Keystone Symposia Boards and Consultants

Ad-Hoc Scientific Advisory Board Reviewers *Participants from the July 2016–June 2017 fiscal year*

Amy L. Brewster, PhD**
Assistant Professor
Behavioral Neuroscience
Purdue University

Nesha S. Burghardt, PhD*
Assistant Professor
Neurobiology
Hunter College, City University of New York

Michael D. Burton, MD, PhD**
Assistant Professor
Systems Neuroscience
University of Texas at Dallas

Preston Campbell, PhD*
Chief Scientific Officer
Monq, Inc.

Genaro Andres Contreras, PhD*
Assistant Professor
Large Animal Clinical Sciences
Michigan State University

David N. Cook, PhD
Executive Vice President of R&D, Chief Scientific Officer
Seres Therapeutics, Inc.

Kizzmekia S. Corbett, PhD**
Research Fellow
Vaccine Research Center
NIAID, National Institutes of Health

Roslyn Nicole Crowder, PhD**
Assistant Professor
Biology
Stetson University

Anne M. Danks, PhD
Vice President
Preclinical Development
Dart NeuroScience LLC

Alison Farrell, PhD
Senior Editor
Nature Medicine
Nature Research

Patrick Finn, PhD
Director
Elpidera
Moderna

Paola Giusti-Rodríguez, PhD*
Research Assistant Professor
Department of Genetics
University of North Carolina at Chapel Hill

Greg Hoffman, PhD
Senior Investigator I
Chemical Biology and Therapeutics
Novartis Institutes for BioMedical Research

Amanda Marie James, PhD*
Assistant Director of IMSD Program
Instructor of Biochemistry
Laney Graduate School, School of Medicine
Department of Biochemistry
Emory University

Satish Jindal, PhD
CEO
Allied Bristol Life Sciences

Michael D.L. Johnson, PhD
Assistant Professor
Department of Immunobiology
University of Arizona

Dallas Jones, PhD
Principal Scientist
Abide Therapeutics

Samira Musah, PhD
Postdoctoral Research Fellow
Wyss Institute
Harvard University

Yava L. Jones-Hall, PhD**
Assistant Professor
Comparative Pathobiology
Purdue University School of Veterinary Medicine

Dominick J. Lemas, PhD*
Assistant Professor
Health Outcomes and Policy
University of Florida College of Medicine

Joseph W. Lewcock, PhD
Head of Biology Discovery
Denali Therapeutics

Rodrigo Maillard, PhD**
Assistant Professor
Department of Chemistry
Georgetown University

Florastina Payton-Stewart, PhD*
Associate Professor of Chemistry
Chemistry
Xavier University of Louisiana

Tom Pohl, PhD
Postdoctoral Fellow
Department of Molecular Biology
Princeton University

Gabriel A. Rabinovich, PhD
Immunopathology
Instituto de Biologia y Medicina Experimental

Michael Rabinowitz, PhD
VP Research
Akebia Therapeutics, Inc.

Miguel A. Sanjuan, PhD
Group Director
Immuno-Oncology Small Molecule Biology
Bristol Myers Squibb Co.

Evan Scott, PhD**
Assistant Professor
Department of Biomedical Engineering
Northwestern University

Veronica A. Segarra, PhD*
Assistant Professor
Biology
High Point University

Glenn E. Simmons, Jr., PhD*
Assistant Professor
Biomedical Sciences
University of Minnesota Medical School Duluth

David R. Soto-Pantoja, PhD**
Department of Cancer Biology
Wake Forest School of Medicine

Kiersten Stead, PhD
Investment Director
Monsanto Growth Ventures
Monsanto Company

Patrick Trojer, PhD
Senior Vice President
EZH2 Franchise and Head
of Translational Sciences Biology
Constellation Pharmaceuticals

Ursula Weiss, PhD
Editorial
Nature Research

Tim Young, PhD
Vertex Pharmaceuticals, Inc.

*2016 Keystone Symposia Fellow
**2017 Keystone Symposia Fellow

2017–2018 Keystone Symposia Conference Series

*Dates, locations and details are subject to possible change. Please visit us online at **www.keystonesymposia.org** or join our various mailing lists or online networks for updates.*

Vectors, Pathogens and Diseases: Current Trends and Emerging Challenges (T1)
Organizers: Maureen Coetzee, Josiane Etang, Stephen Torr and Scott L. O'Neill
Sep 10–14, 2017 | Southern Sun Elangeni & Maharani | Durban, KwaZulu-Natal | South Africa

Maternal-Fetal Crosstalk: Harmony vs. Conflict (T2)
Organizers: Jeff Murray, Louis J. Muglia and Yoel Sadovsky
Oct 4–8, 2017 | Westin Washington, DC City Center | Washington, DC | USA

Regenerative Biology and Applications: Cell Differentiation, Tissue Organization and Biomedical Engineering (T3)
Organizers: Paul K. Tam, Urban Lendahl and Freda D. Miller
Oct 15–19, 2017 | University of Hong Kong | Pok Fu Lam | Hong Kong

Antimicrobials and Resistance: Opportunities and Challenges (T4)
Organizers: Gautam Dantas and Jennifer A. Leeds
Oct 29–Nov 1, 2017 | Eldorado Hotel | Santa Fe, New Mexico | USA

Frontiers of Serotonin Beyond the Brain (T5)
Organizers: Fusun Kilic, Michael D. Gershon and Luc Maroteaux
Nov 12–15, 2017 | Doubletree by Hilton | Park City, Utah | USA

Heart Failure: Crossing the Translational Divide (A1)
Organizers: Yibin Wang, Joseph A. Hill and Carolyn Lam
Jan 14–18, 2018 | Keystone Resort | Keystone, Colorado | USA

State of the Brain: Genetic Dissection of Brain Circuits and Behavior in Health and Disease (A2)
Organizers: Sean Hill, Hongkui Zeng, Z. Josh Huang and György Buzsáki
Jan 14–18, 2018 | Keystone Resort | Keystone, Colorado | USA

T Cell Dysfunction, Cancer and Infection (A3)
Organizers: Daniel C. Douek, W. Nicholas Haining and Jedd D. Wolchok
Jan 16–20, 2018 | Beaver Run Resort | Breckenridge, Colorado | USA

Plant Signaling: Molecular Pathways and Network Integration (A4)
Organizers: Ken Shirasu, Uta Paszkowski and Christian S. Hardtke
Jan 21–24, 2018 | Granlibakken Tahoe | Tahoe City, California | USA

Natural Products and Synthetic Biology: Parts and Pathways (J1)
Organizers: Jon C. Clardy, Yi Tang and Sean F. Brady
Jan 21–24, 2018 | Resort at Squaw Creek | Olympic Valley, California | USA

Tumor Metabolism (A5)
Organizers: Heather Christofk, Christian Metallo and Alec Kimmelman
Jan 21–25, 2018 | Snowbird Resort | Snowbird, Utah | USA

Cell Death, Inflammation and Adaptation to Tissue Stress (A6)
Organizers: Pascal Meier, Eric H. Baehrecke and Kim Newton
Jan 21–25, 2018 | Beaver Run Resort | Breckenridge, Colorado | USA

DNA and RNA Methylation (A7)
Organizers: Chuan He and Ting Wang
Jan 21–25, 2018 | Fairmont Hotel Vancouver | Vancouver, British Columbia | Canada

Organ Crosstalk in Obesity and NAFLD (J3)
Organizers: Gary J. Schwartz, Bei B. Zhang and Christoph Buettner
joint with
Bioenergetics and Metabolic Disease (J4)
Organizers: Patrick Seale, Kendra K. Bence and P. Darrell Neufer
Jan 21–25, 2018 | Keystone Resort | Keystone, Colorado | USA

Ubiquitin Signaling (A8)
Organizers: David Komander and Sylvie Urbé
Jan 28–Feb 1, 2018 | Granlibakken Tahoe | Tahoe City, California | USA

Translational Systems Immunology (A9)
Organizers: Sally John, Soumya Raychaudhuri, Michael Vincent and Mark M. Davis
Jan 28–Feb 1, 2018 | Snowbird Resort | Snowbird, Utah | USA

Precision Genome Editing with Programmable Nucleases (B1)
Organizers: Jin-Soo Kim, Feng Zhang and Daniel F. Voytas
Jan 28–Feb 1, 2018 | Keystone Resort | Keystone, Colorado | USA

Emerging Technologies in Vaccine Discovery and Development (J5)
Organizers: David Kaslow, Nicholas Jackson and Ann L. Lee
joint with
Progress and Pathways Toward an Effective HIV Vaccine (J6)
Organizers: M. Juliana McElrath, Pamela J. Bjorkman and Beatrice H. Hahn
Jan 28–Feb 1, 2018 | Fairmont Banff Springs | Banff, Alberta | Canada

Atherosclerosis: Lessons Learned and Concepts Challenged (B2)
Organizers: Laura F. Michael, Gerard Pasterkamp and Sekar Kathiresan
Feb 4–8, 2018 | Sagebrush Inn & Suites | Taos, New Mexico | USA

Frontiers in Islet Biology and Diabetes (B3)
Organizers: Matthias Hebrok, Seung K. Kim, Felicia Pagliuca and Anil Bhushan
Feb 4–8, 2018 | Keystone Resort | Keystone, Colorado | USA

Cryo-EM from Cells to Molecules: Multi-Scale Visualization of Biological Systems (F1)
Organizers: Georgios Skiniotis, Elizabeth Villa and Andrew B. Ward
Feb 4–8, 2018 | Granlibakken Tahoe | Tahoe City, California | USA

Cancer Epigenetics: New Mechanisms, New Therapies (B4)
Organizers: François Fuks and Anne Brunet
Feb 10–14, 2018 | Beaver Run Resort | Breckenridge, Colorado | USA

Phosphoinositide Biology: New Therapeutic Targets Beyond Class I PI3K (B5)
Organizers: Emilio Hirsch, Tamas Balla and Cristina Donini
Feb 11–15, 2018 | Sagebrush Inn & Suites | Taos, New Mexico | USA

Emerging Cellular Therapies: T Cells and Beyond (B6)
Organizers: Carl H. June, Marcela V. Maus and Bruce R. Blazar
joint with
Lymphocytes and their Roles in Cancer (R1)
Organizers: Iannis Aifantis, Ugur Sahin and Mikala Egeblad
Feb 11–15, 2018 | Keystone Resort | Keystone, Colorado | USA

Mobile Genetic Elements and Genome Plasticity (B7)
Organizers: Marlene Belfort, Evan E. Eichler, Henry L. Levin and Lynne E. Maquat
Feb 11–15, 2018 | Eldorado Hotel | Santa Fe, New Mexico | USA

GPCR Structure and Function: Taking GPCR Drug Development and Discovery to the Next Level (B8)
Organizers: Roger K. Sunahara and Christian Felder
Feb 16–20, 2018 | Eldorado Hotel | Santa Fe, New Mexico | USA

Regulation and Dysregulation of Innate Immunity in Disease (B9)
Organizers: Lynda M. Stuart, Kathryn J. Moore and Kate L. Jeffrey
Feb 18–22, 2018 | Fairmont Hotel Vancouver | Vancouver, British Columbia | Canada

Antibodies as Drugs: Translating Molecules into Treatments (C1)
Organizers: Paul W. H. I. Parren and Erica Ollmann Saphire
Feb 25–Mar 1, 2018 | Whistler Conference Centre | Whistler, British Columbia | Canada

Noncoding RNAs: Form, Function, Physiology (C2)
Organizers: Joshua T. Mendell, Igor Ulitsky and Sohail F. Tavazoie
Feb 25–Mar 1, 2018 | Keystone Resort | Keystone, Colorado | USA

2017–2018 Keystone Symposia Conference Series

*Dates, locations and details are subject to possible change. Please visit us online at **www.keystonesymposia.org** or join our various mailing lists or online networks for updates.*

Endoderm Development and Disease:
Cross-Organ Comparison and Interplay (C3)
Organizers: Xin Sun, Kat Hadjantonakis and Didier Y. R. Stainier
Feb 25–Mar 1, 2018 | Sagebrush Inn & Suites | Taos, New Mexico | USA

Uncomplicating Diabetes:
Reducing the Burden of Diabetes-Related End-Organ Injury (J7)
Organizers: Mark E. Cooper, Thomas M. Coffman, Matthias G. von Herrath and Susan Quaggin
joint with
Vascular Biology and Human Diseases:
From Molecular Pathways to Novel Therapeutics (J8)
Organizers: Elisabetta Dejana, Anne C. Eichmann and Gavin O. Thurston
Feb 25–Mar 1, 2018 | Eldorado Hotel | Santa Fe, New Mexico | USA

Immunological Memory: Innate, Adaptive and Beyond (X1)
Organizers: Rafi Ahmed, Susan M. Kaech and Joseph C. Sun
joint with
Aging, Inflammation and Immunity (X2)
Organizers: Bonnie B. Blomberg and Graham Pawelec
Feb 25–Mar 1, 2018 | Hyatt Regency Austin | Austin, Texas | USA

Manipulation of the Gut Microbiota for Metabolic Health (X3)
Organizers: Nathalie Delzenne and Liping Zhao
joint with
Microbiome, Host Resistance and Disease (X4)
Organizers: Wendy S. Garrett, Yasmine Belkaid and Janelle S. Ayres
Mar 4–8, 2018 | Fairmont Banff Springs | Banff, Alberta | Canada

Cells vs. Pathogens: Intrinsic Defenses and Counterdefenses (C4)
Organizers: Sara R. Cherry, Craig R. Roy and Harmit S. Malik
Mar 23–26, 2018 | Hyatt Regency Monterey | Monterey, California | USA

Cancer Immunotherapy: Combinations (C5)
Organizers: Chris R. Boshoff, Lieping Chen and Lisa Coussens
Mar 23–27, 2018 | Fairmont The Queen Elizabeth | Montreal, Québec | Canada

Chromatin Architecture and Chromosome Organization (X5)
Organizers: Edith Heard and Peter Fraser
joint with
Gene Control in Development and Disease (X6)
Organizers: Richard A. Young, Joanna Wysocka and Phillip A. Sharp
Mar 23–27, 2018 | Whistler Conference Centre | Whistler, British Columbia | Canada

The Resolution of Inflammation in Health and Disease (C6)
Organizers: Catherine Godson, Ira Tabas and Mauro Perretti
Mar 24–28, 2018 | Royal Dublin Society | Dublin | Ireland

iPSCs: A Decade of Progress and Beyond (C7)
Organizers: Shinya Yamanaka, Haruhisa Inoue and Yanhong Shi
Mar 25–29, 2018 | Resort at Squaw Creek | Olympic Valley, California | USA

Organs- and Tissues-on-Chips (D1)
Organizers: Christopher P. Austin, Danilo Tagle, Christine L. Mummery and Brian R. Berridge
Apr 8–12, 2018 | Big Sky Resort | Big Sky, Montana | USA

Myeloid Cells (D2)
Organizers: Edward J. Pearce, Florent Ginhoux and Ana-Maria Lennon-Duménil
Apr 8–12, 2018 | Beaver Run Resort | Breckenridge, Colorado | USA

Therapeutic Targeting of Hypoxia-Sensitive Pathways (V1)
Organizers: Chris W. Pugh, Pablo Wappner, Johanna Myllyharju and Moira K. Whyte
Apr 10–14, 2018 | University of Oxford Mathematical Institute | Oxford | UK

Pushing the Limits of Healthspan and Longevity (D3)
Organizers: Rochelle Buffenstein, Holly Brown-Borg and Colin Selman
Apr 15–19, 2018 | Herrenhausen Palace | Hannover | Germany

Tuberculosis:
Translating Scientific Findings for Clinical and Public Health Impact (X7)
Organizers: Graeme Meintjes, Eric J. Rubin and Sabine Ehrt
joint with
HIV and Co-Infections: Pathogenesis, Inflammation and Persistence (X8)
Organizers: Irini Sereti, Nicolas Chomont and Michaela Müller-Trutwin
Apr 15–19, 2018 | Fairmont Chateau Whistler | Whistler, British Columbia | Canada

Mitochondrial Biology (Z1)
Organizers: Jodi Nunnari, Anu Suomalainen-Wartiovaara and Koji Okamoto
joint with
Selective Autophagy (Z2)
Organizers: Tamotsu Yoshimori, Hong Zhang and Anne Simonsen
Apr 22–26, 2018 | Westin Miyako Kyoto | Kyoto | Japan

Precision Medicine in Cancer (E1)
Organizers: Richard Rosenquist, Elaine Mardis and Charles M. Perou
May 6–10, 2018 | Clarion Hotel Sign | Stockholm | Sweden

Exosomes/Microvesicles:
Heterogeneity, Biogenesis, Function and Therapeutic Developments (E2)
Organizers: Crislyn D'Souza-Schorey and David Lyden
Jun 4–8, 2018 | Beaver Run Resort | Breckenridge, Colorado | USA

One Million Genomes: From Discovery to Health (G1)
Organizers: Geoffrey S. Ginsburg, Teri Manolio and Patrick Boon Ooi Tan
Jun 4–8, 2018 | Herrenhausen Palace | Hannover | Germany

Novel Aspects of Bone Biology (E3)
Organizers: Gerard Karsenty and David T. Scadden
Jun 13–16, 2018 | Snowbird Resort | Snowbird, Utah | USA

B Cells: Mechanisms in Immunity and Autoimmunity (E4)
Organizers: Lars Nitschke, Michael Reth, Roberta Pelanda and David M. Tarlinton
Jun 17–21, 2018 | Maritim Hotel & Conference Center | Dresden | Germany

Advances in Neurodegenerative Diseases Research and Therapy (Z3)
Organizers: Li Gan, Leonard Petrucelli and Morgan H. Sheng
joint with
New Frontiers in Neuroinflammation:
What Happens When CNS and Periphery Meet? (Z4)
Organizers: Marco Prinz, Jonathan Kipnis and Irene Knuesel
Jun 17–21, 2018 | Keystone Resort | Keystone, Colorado | USA

*Conference dates are listed with the first date typically being that of afternoon registration and an evening welcome mixer, and the last day when organized sessions conclude, usually in the evening with a closing plenary session followed by food and entertainment. However, some program formats vary. Please check our website for program specifics for each individual meeting. You can view each program directly by entering **www.keystonesymposia.org** and then **/18** and the alpha-numeric program code (e.g., **www.keystonesymposia/18A1**).*

2018–2019 Keystone Symposia Conference Series

*These meetings are still in development and therefore subject to change. Please visit us online at **www.keystonesymposia.org** to join our mailing list and online networks for updates.*

Drivers of Type 2 Diabetes: From Genes to Environment (S1)
Organizers: Kyong Soo Park, Young-Bum Kim and Zoltan P. Arany
Oct 7–11, 2018 | Grand Hilton Seoul | Seoul | South Korea

Framing the Response to Emerging Virus Infections (S2)
Organizers: Ralph A. Tripp, Suresh Mahalingam, John W. Lowenthal, Andrew G. Bean and Malik Peiris
Oct 14–18, 2018 | University of Hong Kong | Pok Fu Lam | Hong Kong

21st Century Drug Discovery and Development for Global Health (S3)
Organizers: Ken Duncan, Elizabeth Winzeler and Lluís Ballell
Oct 17–20, 2018 | Hotel Palace Berlin | Berlin | Germany

From Rare to Care: Discovery, Modeling and Translation of Rare Diseases (S4)
Organizers: Josef M. Penninger and Kym Boycott
Nov 11–14, 2018 | Vienna BioCenter, IMP Lecture Hall | Vienna | Austria

Leveraging Genomic Diversity to Promote Animal and Human Health (S5)
Organizers: Michèle Ramsay, Han Brunner and Appolinaire Djikeng
Nov 25–29, 2018 | Speke Resort & Conference Centre | Kampala | Uganda

Role of the Genital Tract Microbiome in Sexual and Reproductive Health (S6)
Organizers: Janneke van de Wijgert, Jeanne Marrazzo, Douglas Kwon and Jo-Ann Passmore
Dec 11–15, 2018 | Southern Sun Cape Sun | Cape Town, Western Cape | South Africa

DNA Replication and Genome Instability: From Mechanism to Disease (A1)
Organizers: Karlene A. Cimprich, Mark J. O'Connor and Johannes C. Walter
Jan 13–17, 2019 | Snowbird Resort | Snowbird, Utah | USA

Host and the Environment in IBD:
Scientific Advances Leading to New Therapeutics (A2)
Organizers: Gary D. Wu, Scott Snapper, Judy H. Cho and Aida Habtezion
Jan 13–17, 2019 | Sagebrush Inn & Suites | Taos, New Mexico | USA

Mitochondrial Biology in Heart and Skeletal Muscle (J1)
Organizers: E. Dale Abel and Andrea L. Hevener
joint with
Mitochondria in Aging and Age-Related Disease (J2)
Organizers: Gerald S. Shadel and Holly Van Remmen
Jan 13–17, 2019 | Keystone Resort | Keystone, Colorado | USA

Single Cell Biology (L1)
Organizers: Timm Schroeder and Berthold Göttgens
Jan 13–17, 2019 | Beaver Run Resort | Breckenridge, Colorado | USA

Tuberculosis: Mechanisms, Pathogenesis and Treatment (A3)
Organizers: Christina L. Stallings, Veronique Anne Dartois, Stewart T. Cole and David Barros
Jan 17–21, 2019 | Fairmont Banff Springs | Banff, Alberta | Canada

Integrated Pathways of Disease in NASH and NAFLD (A4)
Organizers: Scott L. Friedman, Arun J. Sanyal, Brent A. Tetri, Mary E. Rinella and Christopher R. Shepard
Jan 20–24, 2019 | Eldorado Hotel | Santa Fe, New Mexico | USA

Digital Health: From Science to Application (A5)
Organizers: Geoffrey S. Ginsburg, Sue Siegel and Eric D. Perakslis
Jan 21–25, 2019 | Keystone Resort | Keystone, Colorado | USA

Windows on the Brain:
Formation and Function of Synapses and Circuits and Disruption in Disease (A6)
Organizers: Kristin Scott, Paola Arlotta, Rui M. Costa and Yimin Zou
Jan 21–25, 2019 | Sagebrush Inn & Suites | Taos, New Mexico | USA

Cellular Plasticity: Reprogramming, Regeneration and Metaplasia (J3)
Organizers: Jason C. Mills, Maike Sander and Ben Z. Stanger
joint with
Signal Dynamics and Signal Integration in Development and Disease (J4)
Organizers: Nicolas Tapon, Liliana Attisano and Raphael Kopan
Jan 27–31, 2019 | Keystone Resort | Keystone, Colorado | USA

Transcription and RNA Regulation in Inflammation and Immunity (B1)
Organizers: Silvia Monticelli, K. Mark Ansel, Sarah Teichmann and Gioacchino Natoli
Feb 2–5, 2019 | Granlibakken Tahoe | Tahoe City, California | USA

Molecular Approaches to Vaccines and Immune Monitoring (J5)
Organizers: Peter D. Kwong, Brandon DeKosky, Giuseppe Ciaramella and Jeffrey B. Ulmer
joint with
B Cell–T Cell Interactions (J6)
Organizers: Gabriel D. Victora and Carola G. Vinuesa
Feb 10–14, 2019 | Keystone Resort | Keystone, Colorado | USA

Obesity and Adipose Tissue Biology (J7)
Organizers: Matthias Blüher, Philipp E. Scherer and Anne Bouloumié
joint with
Functional Neurocircuitry of Feeding and Feeding Disorders (J8)
Organizers: Roger D. Cone, Lori M. Zeltser and Matthew R. Hayes
Feb 10–14, 2019 | Fairmont Banff Springs | Banff, Alberta | Canada

Autophagy: From Model Systems to Therapeutic Opportunities (B2)
Organizers: Vojo Deretic, Li Yu and Leon O. Murphy
Feb 17–21, 2019 | Eldorado Hotel | Santa Fe, New Mexico | USA

Uncovering Mechanisms of Immune-Based Therapy in Cancer and Autoimmunity (B3)
Organizers: Daniel J. Cua, E. John Wherry and Carla V. Rothlin
Feb 18–22, 2019 | Beaver Run Resort | Breckenridge, Colorado | USA

Genome Engineering: From Mechanisms to Therapies (B4)
Organizers: Andrew May, Rodolphe Barrangou and Knut Woltjen
Feb 19–23, 2019 | Fairmont Empress Victoria | Victoria, British Columbia | Canada

Tumor Metabolism (B5)
Organizers: Lewis C. Cantley, Karen H. Vousden and Jeffrey A. Engelman
Feb 24–28, 2019 | Keystone Resort | Keystone, Colorado | USA

Cell Competition in Development and Disease (B6)
Organizers: Margaret A. Goodell, Laura A. Johnston and Thomas P. Zwaka
Feb 24–28, 2019 | Granlibakken Tahoe | Tahoe City, California | USA

Myeloid Cells (B7)
Organizers: Tiffany Horng, Gregory M. Barton and Ajay Chawla
Feb 24–28, 2019 | Eldorado Hotel | Santa Fe, New Mexico | USA

RNA–Protein Interactions (X1)
Organizers: Xiang-Dong Fu, Anne Ephrussi and Douglas L. Black
joint with
Long Noncoding RNAs: From Molecular Mechanisms to Functional Genetics (X2)
Organizers: Saba Valadkhan and Piero Carninci
Feb 24–28, 2019 | Whistler Conference Centre | Whistler, British Columbia | Canada

2018–2019 Keystone Symposia Conference Series

*These meetings are still in development and therefore subject to change. Please visit us online at **www.keystonesymposia.org** to join our mailing list and online networks for updates.*

Phenotypic Drug Discovery:
Recent Advances and Insights from Chemical and Systems Biology (C1)
Organizers: Mark Mercola, Fabien Vincent, Monica Schenone and Arsenio Nueda
Mar 3–7, 2019 | Beaver Run Resort | Breckenridge, Colorado | USA

Diabetes: Innovations, Outcomes and Personalized Therapies (X3)
Organizers: Philip J. Larsen, Melina Claussnitzer and Tina Vilsbøll
joint with
Unraveling the Secrets of Kidney Disease (X4)
Organizers: Katalin Susztak, Caroline Fox and Masaomi Nangaku
Mar 3–7, 2019 | Whistler Conference Centre | Whistler, British Columbia | Canada

Cancer Immunotherapy: Mechanistic Insights to Improve Clinical Benefit (C2)
Organizers: Padmanee Sharma, Aviv Regev, Crystal L. Mackall and Kristen Hege
Mar 10–14, 2019 | Whistler Conference Centre | Whistler, British Columbia | Canada

Microbiome: Chemical Mechanisms and Biological Consequences (C3)
Organizers: Dennis W. Wolan, Peter J. Turnbaugh and Emily P. Balskus
Mar 10–14, 2019 | Fairmont The Queen Elizabeth | Montréal, Québec | Canada

Innate Immune Receptors: Roles in Immunology and Beyond (M1)
Organizers: Jenny P.Y. Ting, Shie-Liang Edmond Hsieh, Fu-Tong Liu, Michael Gale, Jr. and Siamon Gordon
Mar 10–14, 2019 | International Conference Hall of Academia Sinica | Taipei | Taiwan

Mammalian Sensory Systems (C4)
Organizers: Stephen Liberles, David D. Ginty, Jeffrey R. Holt and Melanie Samuel
Mar 15–19, 2019 | Allen Institute | Seattle, Washington | USA

Cancer Metastasis:
The Role of Metabolism, Immunity and the Microenvironment (M2)
Organizers: Erika L. Pearce, Sarah-Maria Fendt, Russell G. Jones and Peter F. Carmeliet
Mar 15–19, 2019 | Firenze Fiera – Fortezza da Basso | Florence | Italy

Epigenetics and Human Disease (X5)
Organizers: Cheryl H. Arrowsmith, Anne Schaefer and Mark A. Dawson
joint with
3D Genome: Gene Regulation and Disease (X6)
Organizers: Bing Ren, Ana Pombo and Joseph R. Ecker
Mar 17–21, 2019 | Fairmont Banff Springs | Banff, Alberta | Canada

Origins of Allergic Disease: Microbial, Epithelial and Immune Interactions (M3)
Organizers: Marsha Wills-Karp, Donald Leung and Kari Nadeau
Mar 24–27, 2019 | Granlibakken Tahoe | Tahoe City, California | USA

Innate and Non-Classical Immune Cells in Cancer Immunotherapy (C5)
Organizers: Nicholas D. Huntington, Eric Vivier, Caroline Robert and Lewis L. Lanier
Mar 24–28, 2019 | Keystone Resort | Keystone, Colorado | USA

HIV Vaccines (X7)
Organizers: Nicole A. Doria-Rose, Thumbi Ndung'u and Gunilla Karlsson Hedestam
joint with
Functional Cures and the Eradication of HIV (X8)
Organizers: Lynn Morris, Melanie M. Ott and Kevin V. Morris
Mar 24–28, 2019 | Fairmont Chateau Whistler | Whistler, British Columbia | Canada

Lipidomics and Functional Metabolic Pathways in Disease (C6)
Organizers: Sarah Spiegel, Charles N. Serhan and Valerie B. O'Donnell
Mar 31–Apr 4, 2019 | Steamboat Grand | Steamboat Springs, Colorado | USA

Leveraging the Revolution in Resolution: Imaging Across Scales (D1)
Organizers: Bo Huang and Elizabeth R. Wright
Apr 7–10, 2019 | Snowbird Resort | Snowbird, Utah | USA

Antibodies as Drugs:
New Horizons in the Therapeutic Use of Engineered Antibodies (D2)
Organizers: Christian Klein, Mark S. Cragg and Germaine Fuh
Apr 7–11, 2019 | Beaver Run Resort | Breckenridge, Colorado | USA

Proteomics and its Application to Translational and Precision Medicine (D3)
Organizers: Matthias Mann, Emma K. Lundberg, Albert J.R. Heck and Mathias Uhlén
Apr 7–11, 2019 | Clarion Hotel Sign | Stockholm | Sweden

Immune, Epithelial and Microbiome Crosstalk in Skin Health and Disease (D4)
Organizers: Michel Gilliet, Emma Guttman, Anthony Oro and Manolis Pasparakis
Apr 8–11, 2019 | Herrenhausen Palace | Hannover | Germany

Biomolecular Condensates:
Phase-Separated Organizers of Cellular Biochemistry (D5)
Organizers: Michael K. Rosen and Anthony Hyman
Apr 10–13, 2019 | Snowbird Resort | Snowbird, Utah | USA

Immunometabolism, Metaflammation and Metabolic Disorders (D6)
Organizers: Gökhan S. Hotamisligil, Ruslan M. Medzhitov and Karine Clément
Apr 14–18, 2019 | Fairmont Hotel Vancouver | Vancouver, British Columbia | Canada

Small Regulatory RNAs (D7)
Organizers: V. Narry Kim, Gregory J. Hannon and Lin He
Apr 14–18, 2019 | Daejeon Convention Center | Daejeon | South Korea

Delivering Therapeutics Across Biological Barriers (E1)
Organizers: David Brayden, Claus-Michael Lehr and Kathryn Whitehead
May 6–9, 2019 | Royal Dublin Society | Dublin | Ireland

Climate Change-Linked Stress Tolerance in Plants (M4)
Organizers: Julian Schroeder and Julia Bailey-Serres
May 13–16, 2019 | Herrenhausen Palace | Hannover | Germany

Positive-Strand RNA Viruses (E2)
Organizers: Frank van Kuppeveld and Andrea Gamarnik
Jun 9–13, 2019 | INEC–Ireland's National Events Centre | Killarney, Co. Kerry | Ireland

Neural Environment in Disease: Glial Responses and Neuroinflammation (Z1)
Organizers: Richard Daneman, Dorothy Schafer, Michael V. Sofroniew and Vanda A. Lennon
joint with
Neurodegenerative Diseases: New Insights and Therapeutic Opportunities (Z2)
Organizers: Valina L. Dawson, Joseph W. Lewcock and Fred H. Gage
Jun 16–20, 2019 | Keystone Resort | Keystone, Colorado | USA

*Conference dates are listed with the first date typically being that of afternoon registration and an evening welcome mixer, and the last day when organized sessions conclude, usually in the evening with a closing plenary session followed by food and entertainment. However, some program formats vary. Please check our website for program specifics for each individual meeting. You can view each program directly by entering **www.keystonesymposia.org** and then **/19** and the alpha-numeric program code (e.g., **www.keystonesymposia/19A1**).*

Keystone Symposia - Speaker Abstracts
Cryo-EM from Cells to Molecules: Multi-Scale Visualization of Biological Systems (F1)

Current as of 04/13/2018 09:04

Only the abstract text is displayed here, not the pictures or graphs.

De novo Design of Self-Assembling Protein Nano Materials

David Baker
Institute for Protein Design and HHMI, University of Washington, Seattle, WA, USA

I will describe the computational design, from de novo designed protein building blocks, of self-assembling homo-oligomers with cyclic and dihedral symmetries, multicomponent assemblies with tetrahedral, octahedral, and icosahedral symmetries, unbounded one dimensional fibers with a range of helical symmetries, and multicomponent two dimensional layers with wall paper group symmetries. X-ray crystallography and Cryo-EM have shown these nano materials have rigid well defined structures almost identical to the computational design models, and I will discuss the use of these stable, rigid and readily modifiable scaffolds for cryoEM structure determination. I will also describe the de novo design and high resolution structure determination of complex multi-pass membrane proteins, and the elements of a new de novo designed protein logic based on modular buried hydrogen bond networks: large sets of mutually orthogonal heterodimers and heterotrimers, pH sensitive protein switches, and allosterically controlled protein binding.

1

Architectures of Lipid Transport Systems in Bacteria

Gira Bhabha
New York University School of Medicine, Skirball Institute, New York, NY, USA

The outer membrane of double membraned bacteria is the first line of defense for these bacteria against environmental stresses. It is an asymmetric lipid bilayer, comprising LPS on the outer leaflet and phospholipids on the inner leaflet. The mechanism of LPS transport has been relatively well resolved in recent years. However, the fundamental question of how phospholipids are trafficked between inner and outer membranes remains mysterious. We used a structure-driven approach to studying a family of proteins called the MCE (Mammalian Cell Entry) proteins in E. coli, some of which had previously been implicated in phospholipid transport. Our results reveal the architectures, conformational dynamics and potential mechanisms for how MCE proteins may transport lipids between membranes in double membraned bacteria.

2

Sample Vitrification Issues and Some Ideas about How to Overcome Them

Bridget Carragher[1], Venkata Dandey[1], Hui Wei[2], Zhening Zhang[1], Alex Noble, Clinton S. Potter
[1] The National Resource for Automated Molecular Microscopy, Simons Electron Microscopy Center, the New York Structural Biology Center, New York, NY, USA

We are developing a novel device for vitrifying samples for cryo-electron microscopy (cryoEM). The device, called Spotiton, uses a piezo electric dispensing head to deliver a series of droplets to a “self-blotting” grid composed of nanowires surrounding the substrate that effectively act as blotting paper to whisk the sample from the substrate, leaving behind a thin wet film that is then plunged into liquid ethane for rapid vitrification. We have used Spotiton and the nanowire grids to prepare several test specimens that can be reconstructed to sub-3A resolution, indicating that the process has no apparent deleterious effect on the sample integrity. The system is now in routine and continuous use in our lab and has been used to successfully vitrify a wide variety of samples including small and large soluble proteins, integral membrane proteins, samples with and without detergents, in nanodiscs and in the presence of glycerol. We are also exploring whether this new approach can be useful in alleviating some of the deleterious effects of the air-water interface.

Funding: Supported by NIH National Institute of General Medical Sciences (GM103310) and the Simons Foundation (349247).

3

Cryo-EM as a Tool for Structure Determination in Industry

Tom Ceska
UCB Celltech, Slough, UK

Cryo-EM is an emerging disruptive technology. It is complementary to X-ray crystallography and NMR, with equivalent resolution now possible. It is applicable when crystallography cannot deliver structures and to membrane proteins which have proved to be intractable. The key feature of cryo-EM structures is that the molecules are trapped in solution, and are unconstrained.

Applications include antibody and small molecule drug discovery to influence design of therapeutic candidates in real time. This includes epitope mapping, small molecule location and orientation and conformational definition of a target.

UCB is participating in a consortium with 4 other companies investigating single particle cryo-EM for use in drug discovery. The microscope, a Titan Krios, has been installed at the Nanoscience Centre in Cambridge.

With new higher performance electron microscopes, better software and direct electron detectors, it is now possible to capture structural information at close to atomic resolution.

I will describe some of the issues we discovered we had to overcome, and a couple of early examples from data sets we have collected.

4

Integrated Structural Biology of Gene Transcription

Patrick Cramer
Max Planck Institute for Biophysical Chemistry, Göttingen, Germany

Our laboratory combines integrated structural biology with functional genomics and computational biology to study the mechanisms of gene transcription and its regulation in eukaryotic cells. Recent work includes structural studies of transcription initiation and regulation (Kohler Science 2017; Engel Cell 2017; Nozawa Nature 2017; Schilbach Nature 2017; Farnung Nature 2017) and the development of transient transcriptome sequencing (TT-seq), which monitors changes in enhancer landscapes at high temporal resolution (Schwalb Science 2016; Demel Mol. Syst. Biol. 2017). I will present our latest structural studies, which have used a combination of cryo-electron microscopy, X-ray crystallography and crosslinking, to arrive at the structure of the RNA polymerase II transcription pre-initiation complex. This complex comprises 46 proteins and has a molecular weight of around 2 MegaDalton. I will concentrate on the critical steps in this structure solution, which also involved methods development. Further I will discuss our most recent structural study that reveals a chromatin remodeler bound to the nucleosome and explains how Snf2-type remodelers are functionally controlled. These results provide a framework for understanding how transcription in eukaryotic cells is achieved.

5

Single Particle Imaging with the Volta Phase Plate

Radostin Danev, Maryam Khoshouei, Wolfgang Baumeister

Max Planck Institute of Biochemistry, Martinsried, Germany

The Volta phase plate (VPP) is a remarkably simple device comprising a thin (~10 nm) amorphous carbon film positioned at the back focal plane of the objective lens [1]. The film is continuously heated at ~250 °C to prevent beam-induced contamination and enable the Volta potential effect. The strong central diffraction beam of unscattered electrons interacts with the film and modifies its surface properties leading to the creation of a Volta potential difference (the outer potential just above the film surface) between the central spot and the surrounding areas. This in turn leads to a three-dimensional electrostatic potential distribution above and below the film which when integrated along the beam path results in a phase shift difference between the central beam and the scattered beams. The effect of the central diffraction beam on the film is cumulative, producing a phase shift which increases with the electron dose. The increase is not linear and slows down as the phase shift approaches ~π/2 [1], which coincidently is the ideal phase shift for a phase plate.

Single particle applications of the VPP are still in their early days, but the results thus far are very encouraging. The improved contrast was instrumental in the near-atomic cryo-EM structure determination of peroxiredoxin-3 (~250 kDa) [2], the nucleosome (~200 kDa) [3], a calcitonin class-B GPCR (~150 kDa) [4] and hemoglobin (64 kDa) [5]. The VPP has the capability to extend the applicability of cryo-EM towards smaller particles and it is not unreasonable to expect that in the near future even structures under 50 kDa could be solved to near-atomic resolutions.

[1] R Danev et al, PNAS 111 (2014) p. 15635.
[2] M Khoshouei et al, Nature Communications 7 (2016) a.n. 10534.
[3] E Y D Chua et al, Nucleic Acids Research 44 (2016) p. 8013.
[4] Y L Liang et al, Nature 546 (2017) p. 118.
[5] M Khoshouei et al, Nature Communications 8 (2017) a.n. 16099.

6

Revealing Secrets Hiding in Plain Sight

Mark H. Ellisman, Ph.D.
Distinguished Professor of Neurosciences and Bioengineering, Director, National Center for Microscopy and Imaging Research (NCMIR), University of California, San Diego, Senior Fellow, HHMI Janelia Research Campus, La Jolla, CA, USA

A grand goal of biomedical and brain research is to understand how the interplay of structural, chemical and electrical signals in and between all cells in their tissue environment give rise to normal and abnormal functioning. New technologies are hastening progress as biologists make use of an increasingly powerful arsenal of tools and technologies for obtaining data, from the level of molecules to whole organs. This talk will highlight projects in which development and application of new contrasting methods and 3D microscopic imaging tools have allowed us to observe otherwise complex or hidden relationships between cellular, subcellular and molecular constituents of cells and tissues.

7

Structural Basis of Mitochondrial Receptor Binding and GTP-Driven Conformational Constriction by Dynamin-Related Protein 1

Raghav Kalia[1,2,3,5], Ray Yu-Ruei Wang[1,3,4], Ali Yusuf[1,3], Paul V. Thomas[1,3,5], David A. Agard[1,3,4], Janet M. Shaw[2,4], Adam Frost[1,2,3,5]
[1]Department of Biochemistry and Biophysics, University of California, San Francisco, San Francisco, CA, USA;
[2]Department of Biochemistry, University of Utah, Salt Lake City, UT, USA
[3]California Institute for Quantitative Biomedical Research, San Francisco, CA, USA
[4]Howard Hughes Medical Institute; [5]Chan Zuckerberg Biohub, San Francisco, CA, USA
Email: adam.frost@ucsf.edu

Mitochondrial inheritance, genome maintenance, and metabolic adaptation all depend on organelle fission by Dynamin-Related Protein 1 (DRP1) and its mitochondrial receptors. DRP1 receptors include the paralogs Mitochondrial Dynamics 49 and 51 (MID49/MID51) and Mitochondrial Fission Factor (MFF), but the mechanisms by which these proteins recruit DRP1 and regulate its activities are unknown. Here we present a cryoEM structure of human, full-length DRP1 bound to MID49 and an analysis of structure- and disease-based mutations. We report that GTP binding allosterically induces a remarkable elongation and rotation of the G-domain, Bundle-Signaling Element (BSE) and connecting hinge loops of DRP1. In this nucleotide-bound conformation, a distributed network of multivalent interactions promotes DRP1 copolymerization into a linear filament with MID49, MID51 or both. Subsequent GTP hydrolysis and exchange within the filament leads to receptor dissociation, shortening through disassembly, and concomitant curling of DRP1 oligomers into closed rings. The dimensions of the closed DRP1 rings are consistent with DRP1-constricted mitochondrial tubules observed in human cells. These structures are the first views of full-length, receptor- and nucleotide-bound dynamin-family GTPases and—in comparison with nucleotide-free crystal structures—teach us how these molecular machines perform mechanical work through nucleotide-driven allostery.

8

Structure and Mechanism of AMPA Receptor - Auxiliary Protein Complexes

Eric Gouaux
Vollum Institute, Oregon Health & Science University, Portland, OR, USA

AMPA receptors mediate fast excitatory neurotransmission in the mammalian brain and transduce the binding of presynaptically released glutamate to the opening of a transmembrane cation channel. Within the postsynaptic density, however, AMPA receptors coassemble with transmembrane AMPA receptor regulatory proteins (TARPs), yielding a receptor complex with altered gating kinetics, pharmacology, and pore properties. I will present cryo-EM structures of the GluA2-TARP γ2 complex in the presence of the partial agonist kainate or the full agonist quisqualate together with a positive allosteric modulator or with quisqualate alone. Our structural studies show how TARPs sculpt the ligand-binding domain gating ring, enhancing kainate potency and diminishing the ensemble of desensitized states. TARPs encircle the receptor ion channel, stabilizing M2 helices and pore loops, illustrating how TARPs alter receptor pore properties. Structural and additional computational analysis suggests the full agonist and modulator complex harbors an ion-permeable channel gate, providing the first view of an activated AMPA receptor.

9

Applications of Cryo-EM in Small Molecule and Biologics Drug Design

Seungil Han
Medicine Design, Pfizer Inc, Groton, CT, USA

Since the introduction of direct electron detectors, the resolution and range of biological molecules amenable to single particle cryo-EM have significantly widened. The prospects of studying protein-ligand interactions of large macromolecular complexes such as ribosome, viral glycoprotein complexes, ion-channels, gamma-secretase etc. without having to generate single crystal, are definitely appealing. Since the grand opening of our state of art facility in October, 2016, we have started to work on several targets to support the discovery of new drugs and vaccines. I will describe the applications of cryo-EM and progresses we have made. We expect many more examples to follow as the technique is applied to a broad range of clinical targets that remain inaccessible to techniques such as X-ray crystallography, providing valuable new insights for future structure based drug design.

10

CryoEM Structure of Mammalian Respiratory Complex I, an Asymmetric 1 MDa Energy-Converting Enzyme

Judy Hirst
MRC Mitochondrial Biology Unit, University of Cambridge, Cambridge, UK

Respiratory complex I is one of the largest membrane-bound enzymes in the mammalian cell. It powers ATP synthesis in mitochondria by capturing the free energy produced by electron transfer from NADH to ubiquinone and using it to drive protons across the inner membrane. Mammalian complex I contains 45 subunits. 14 core subunits house the catalytic machinery and are conserved from bacteria to humans, whereas the cohort of 31 supernumerary subunits forms an exoskeleton around the core and is specific to mammals species. In this talk I will present our studies of the structure of mammalian complex I (NADH:ubiquinone oxidoreductase) by single-particle cryoEM, including how all 45 subunits were located and modelled in the density map, and results from computational sorting of the particles that led to identification of different structural classes. The classes are related by subtle domain movements and reveal conformationally-dynamic regions that are relevant to catalysis and to enzyme state transitions that occur during hypoxia and ischaemia.

11

Structure and Function of the Bacterial Type IV Secretion System through Electron Cryotomography

Grant Jensen
Caltech, HHMI, Pasadena, CA, USA

In the last several years electron cryotomography (ECT) has made it possible to visualize large macromolecular assemblies inside intact cells in a near-native, "frozen-hydrated" state in 3-D to a few nanometers resolution. Increasingly, atomic models of individual proteins and smaller complexes obtained by X-ray crystallography, NMR spectroscopy, or single particle cryo-EM can be fit into cryotomograms to reveal how the various pieces work together inside cells. In this talk I will present our recent ECT work on the bacterial type IV secretion system (T4SS). The T4SS is an amazingly versatile nanomachine involved in exchange of DNA between bacteria (conjugation, the basis for modern biotechnology), DNA uptake from the extracellular environment, DNA secretion into host cells (the basis of modern plant genetics), and the deposition of hundreds of different protein effector molecules into hosts during infection (including multiple human diseases). We have resolved its basic structure in several model systems and been able to pinpoint the locations of most of the major components. Excitingly, ECT has also allowed us to visualize membrane tubes used by the T4SS to transport cargo to other cells.

12

How Low Can You Go? Size and Resolution Limits using Conventional Cryo-EM

Gabriel C. Lander
The Scripps Research Institute, Integrative Structural and Computational Biology, La Jolla, CA, USA

Advances in single particle cryoEM have enabled determination of macromolecular structures at resolutions previously only attainable by X-ray crystallography. However, nearly 98% of the structures resolved to better than 4 Å resolution have been determined using top-of-the-line 300 keV transmission electron microscopes. We show that it is possible to obtain cryo-EM reconstructions of macromolecular complexes at a range of sizes to better than 3 Å resolution using a mid-range 200 keV TEM with a K2 detector. These structures are of sufficient quality to unambiguously assign amino acid rotameric conformations and identify ordered water molecules and bound co-factors, features previously thought only to be resolvable using TEMs operating at 300 keV.

13

Single Particle Cryo-EM Studies of Lipid-Protein Interactions

Wei Mi[1], Yanyan Li[2], Sung Hwan Yoon[3], Robert K. Ernst[3], Thomas Walz[4], Maofu Liao[1]
[1]Department of Cell Biology, Harvard Medical School, Boston, Massachusetts, USA; [2]State Key Laboratory of Food Science and Technology, Jiangnan University, Wuxi, China; [3]Department of Microbial Pathogenesis, School of Dentistry, University of Maryland, Baltimore, Maryland, USA;
[4]Laboratory of Molecular Electron Microscopy, the Rockefeller University, New York, New York, USA

The transbilayer movement of most lipids is energetically unfavorable and requires the facilitation by lipid flippases. The mechanisms by which flippases recognize and translocate specific lipid molecules have remained elusive. Lipopolysaccharide (LPS) in the outer membrane of Gram-negative bacteria is critical for their cell envelope assembly. LPS synthesized in the cytoplasmic leaflet of the inner membrane is flipped to the periplasmic leaflet by MsbA, an ATP-binding cassette transporter. We have used single-particle cryo-EM to elucidate the structures of lipid nanodisc-embedded MsbA in three functional states. Our study uncovers the long-sought-after structural basis for LPS recognition by MsbA, delineates the conformational transitions of MsbA to flip LPS, and paves the way for structural characterization of other lipid flippases.

14

High-Throughput Cryo-Electron Tomography: Visualizing Host-Pathogen Interactions at High Resolution

Jun Liu
Yale School of Medicine, Microbial Pathogenesis, West Haven, CT, USA

To establish and maintain a successful infection, microbial pathogens have evolved a variety of strategies to invade the host and establish a complex host-pathogen crosstalk that leads to subversion of cellular functions and establishment of diseases. Our understanding of bacterial pathogenesis is therefore dominated largely by cell biology of the complex host-pathogen interactions. In my talk, cryo-electron tomography is not only utilized to visualize bacterial pathogens at high resolution, but also tuned specifically for detailed characterization of host-pathogen interactions.

15

High-Resolution Cryo-EM Studies of Retroviral Intasomes

Dmitry Lyumkis[1*], Dario Passos[1], Li Min[2], Renbin Yang[2], Stephanie Rebensburg[3], Rodolfo Ghirlando[2], Youngmin Jeon[1], Nikoloz Shkriabai[3], Mamuka Kvaratskhelia[3], Robert Craigie[2]
[1]The Salk Institute for Biological Studies, Laboratory of Genetics, La Jolla, CA, USA; [2]National Institutes of Health, National Institute of Diabetes and Digestive Diseases, Bethesda, MD, USA; [3]Ohio State University, Center for Retrovirus Research and College of Pharmacy, Columbus, OH, USA
*Email: dlyumkis@salk.edu

Like all retroviruses, HIV-1 irreversibly inserts a viral DNA (vDNA) copy of its RNA genome into host target DNA (tDNA). The intasome, a higher-order nucleoprotein complex composed of viral integrase (IN) and the ends of linear vDNA, mediates integration. Productive integration into host chromatin initiates with the formation of the soluble synaptic complex, an assembly of IN around vDNA oligonucleotides, and then proceeds through a multi-step reaction mechanism to result in the formation of the strand transfer complex (STC) containing catalytically joined vDNA and tDNA. HIV-1 intasomes have been refractory to high-resolution structural studies, despite considerable effort. We have previously used a mutant IN enzyme to produce soluble HIV STCs, which provided the first high-resolution structure of an HIV intasome and highlighted how HIV-1 can use the common retroviral intasome core architecture to accommodate different IN domain modules for assembly. I will discuss progress and current challenges pertaining to high-resolution cryo-EM studies of HIV intasomes.

This work is supported by:
NIH 2U54GM103368
NIH 1R01AI136680

16

Single Particle Cryo-EM of a 200 kDa Magnesium Ion Channel

Doreen Matthies[1], Olivier Dalmas[2], Alan Merk[1], Bharat G. Reddy[2], Alberto Bartesaghi[1], Mario Borgnia[1], Eduardo Perozo[2*], Sriram Subramaniam[1*]
[1]Laboratory of Cell Biology, Center for Cancer Research, National Cancer Institute, NIH, Bethesda, MD, USA;
[2]Department of Biochemistry and Molecular Biology, Institute for Biophysical Dynamics, The University of Chicago, Chicago, IL, USA
*Corresponding Author

CorA, the major Mg^{2+}-uptake system in prokaryotes, is a 200 kDa homo-pentameric channel and gated by intracellular Mg^{2+}. X-ray crystallographic studies of CorA show similar conformations under Mg^{2+}-bound and Mg^{2+}-free conditions, but EPR spectroscopic studies reveal large Mg^{2+}-driven quaternary conformational changes. Major developments in the field of cryo-EM in the last few years allowed to finally get a glimpse of the structure of the open conformation by single-particle cryo-EM. After optimization of various parameters such as protein purification, protein concentration, buffer conditions, EM-grid pre-treatment, and blotting conditions, we were able to determine structures of CorA in the Mg^{2+}-bound closed conformation and in two open Mg^{2+}-free states at resolutions of 3.8, 7.1, and 7.1 Å, respectively. In the absence of bound Mg^{2+}, four of the five subunits are displaced to variable extents (10-25 Å) by hinge-like motions as large as 35° at the stalk helix. The transition between a single 5-fold symmetric closed state and an ensemble of low Mg^{2+}, open, asymmetric conformational states is, thus, the key structural signature of CorA gating.

Matthies, D., Dalmas, O., Borgnia, M.J., Dominik, P.K., Merk, A., Rao, P., Reddy, B.G., Islam, S., Bartesaghi, A., Perozo, E., et al. (2016). Cryo-EM Structures of the Magnesium Channel CorA Reveal Symmetry Break upon Gating. Cell 164, 747-756.

17

Cryo-EM Snapshots of the Spliceosome Provide Insights into the Mechanism of pre-mRNA Splicing

Kiyoshi Nagai
MRC Laboratory of Molecular Biology, Francis Crick Avenue, Cambridge, UK

In the past two years we have determined near atomic resolution structures of yeast spliceosomes in several key intermediate states (1-5). These structures have begun to reveal the detailed mechanical working of this intricate molecular machine (1). The yeast pre-catalytic B complex structure (2) shows that U2 snRNP is held over the tri-snRNP (3) and the branch helix is bound to SF3b. The single stranded region of U4 snRNA is already bound to the active site of the Brr2 helicase ready to unwind the U4/U6 snRNA duplex for activation. In the C complex immediately after the first trans-esterification reaction (4) the cleaved 5'exon is held between the Large and N-terminal domains of Prp8 and the 5'-phosphate of the first intron nucleotide (G1) linked to the branch point adenosine is positioned at the catalytic center. The branch helix is docked into the active site by step 1 factors. The Prp16 helicase is poised to release the step 1 factors and the branch helix from the active site. The C* complex structure after Prp16 action (5) shows that the dissociation of step 1 factors from the active site induce conformational change for the incoming 3'exon binding. The RNA-based catalytic core of the spliceosome formed during the B to Bact transition remains unchanged throughout the catalytic phase of the splicing. The DEAH-box RNA helicases and step specific factors regulate docking and undocking of the substrates (branch site and 3' splice site) to the single RNA-based active site to catalyze both branching and exon-ligation reactions.

References

1. Fica SM; Nagai K. **Nature Struc. Molec. Biol.** (2017) in press.
2. Plaschka C; Lin, P-C; Nagai K. **Nature** 546, 617 (2017).
3. Nguyen TH; Galej WP; Bai XC; Oubridge C; Newman AJ; Scheres SH; Nagai K. **Nature** 530, 298 (2016).
4. Galej WP; Wilkinson ME; Fica SM; Oubridge C; Newman AJ; Nagai K. **Nature** 537, 197 (2016).
5. Fica SM; Oubridge C; Galej WP; Wilkinson WE; Bai XC; Newman AJ; Nagai K. **Nature** 542, 377 (2017).

18

Probing the Molecular Organization of Cells and Organelles using Cryo-Electron Microscopy

Daniela Nicastro
Departments of Cell Biology and Biophysics, University of Texas Southwestern Medical Center, Dallas, TX, USA

Rapid freezing of cells can provide outstanding structure preservation and time resolution of dynamic cellular processes. Electron tomography of rapidly frozen specimens (cryo-ET) is a powerful technique for imaging biological structures in their native state and in an unperturbed cellular environment. We integrate high resolution imaging by either cryo-ET and sub-tomogram averaging or TYGRESS (Tomography-Guided 3D Reconstruction of Subcellular Structures), with comparative genetics, biochemical methods and EM-visible labeling to deconstruct the in situ 3D structure and functional organization of macromolecular complexes. Among different model systems, we use e.g. cilia and flagella to advance techniques and approaches for high-resolution imaging of complex cellular structures. Cilia and flagella are conserved and ubiquitous eukaryotic organelles that are composed of more than 600 different proteins and have important biological roles in motility and sensation; defects in their assembly or function cause severe human diseases. Our cryo-ET studies visualize the three-dimensional structures of intact wild-type and mutant flagella, and dissect the organization of key macromolecular complexes in different functional states. Such information can provide detailed insights into the structural basis and ultimately the function of many cellular processes.

19

Challenges and Progress in the Study of Large Human Complexes in Gene Expression Regulation

Avinash Patel[1], Simon Poepsel[1], Vignesh Kasinath[1], Robert Louder[1], Jie Fang[2], Eva Nogales[1,2,3]
[1] University of California, Berkeley, Berkeley, CA, USA; [2] Howard Hughes Medical Institute
[3] Lawrence Berkeley National Lab, Berkeley, CA, USA

Cryo-EM allows the structural characterization of macromolecules without the need for crystallization. But while for certain type of samples that are ideally suited for cryo-EM studies it has been possible to reach high-resolution structures following relatively standard procedures, other biological systems remain highly challenging due to the scarcity of the sample, poor stability of the complexes, and most often, their intrinsic flexibility. Among these challenging samples are eukaryotic complexes involved in the essential task of regulating gene expression. I will present our efforts to overcome the challenges we face in the study of human transcription initiation complexes and the epigenetic gene silencer Polycomb repressive complex 2 (PRC2).

Funding: HHMI, NIGMS, Eli Lilly

20

Integrative Modeling of Biomolecular Assembly Structures and Pathways

Andrej Sali
University of California, San Francisco, Bioengineering and Therapeutic Sciences, San Francisco, CA, USA

The networks and spatial structures of biomolecular interactions provide insights into their function and thus help us to understand the workings of living cells. Detailed structural characterization of large and often dynamic assemblies and their networks is generally impossible by any single existing experimental or computational method. This challenge can be overcome by hybrid approaches that integrate data from diverse biophysical experiments (eg, X-ray crystallography, NMR spectroscopy, electron microscopy, chemical cross-linking, yeast-two hybrid system, and various chemical genetics and proteomics approaches). We formulate the hybrid approach to structure and/or network determination as an optimization problem, the solution of which requires three main components: the representation of the assembly or network, the scoring function, and the optimization method. The ensemble of solutions to the optimization problem embodies the most accurate characterization given the available information. The key challenges remain translating experimental data into restraints on the structure and/or network, combining these spatial and/or network restraints into a single scoring function, optimizing the scoring function, and analyzing the resulting ensemble of solutions. The approach will be illustrated by several applications to specific biological systems, including the structure determination of the nuclear pore complex and the mapping of the gulonate pathway in *Haemophilus influenzae.*

21

Structure-Based Vaccine Design

Christopher Cottrell, Gabriel Ozorowski, Hannah Turner, Bartek Nogal, Kimmo Rantalainen, Zachary Berndsen, Sasha Moola, Matteo Bianchi, Jeong Hyun Lee, Elise Landais, Pascal Poignard, Ian Wilson, Lars Hangartner, Andrew B. Ward
The Scripps Research Institute, La Jolla, CA, USA

Rational, structure-based vaccine design, which is being pursued against HIV, requires intimate knowledge of antigen-antibody interactions. CryoEM has now proven to be a critical tool for providing such details, with distinct advantages in throughput, sample requirement, and ability to analyze complex and heterogeneous samples. For HIV, the antigenic target for vaccines is the envelope glycoprotein trimer on the surface of the virion because it is the target for all broadly neutralizing antibodies (bnAbs). Various soluble Env trimer reagents that structurally and antigenically mimic native Env have now been deployed and tested in animal immunization studies. We use these soluble Env trimers form complexes and solve structures of broadly neutralizing antibody (bnAb) lineages from infected patients, as well as neutralizing and non-neutralizing antibody complexes elicited via vaccination. In solving these complex structures, we aim to understand the basis of productive immune responses in humans and compare them to vaccination experiments in different animal models. Because of the complexity of both Env and polyclonal immune responses to large, glycosylated antigens we are currently working to improve methods and throughput. The structural information that we generate can subsequently be used for immunogen design and re-design, with the goal of programming immunity in a rational manner. We are also using EM and other biophysical methods to help shepherd Env trimer vaccine candidates toward the clinic for human testing.

This work was supported by NIH grants Scripps CHAVI-ID (UM1 AI100663), P01 AI110657, International AIDS Vaccine Initiative, and the Bill and Melinda Gates Foundation CAVD (OPP1115782 and OPP1084519)

22

Cryo-EM: A New Tool for Drug Discovery at Novartis

Christian Wiesmann
Novartis Institutes for Biomedical Research, Chemical Biology and Therapeutics, Basel, Switzerland

The advent of the resolution revolution established single particle cryo electron microscopy as a serious contender in structural biology for the investigation of complex biological systems. Novartis, in collaboration with the Friedrich Miescher Institute, invested in a state-of-the-art EM facility at our research site in Basel with the goal to harness the potential of this method. In addition to investigating large protein complexes, we strive to enable structure-based-drug discovery on systems that are very difficult or even inaccessible to crystallographic studies. I will present the progress we have made since the opening of our EM-center in late 2016 and discuss the impact cryo-EM has had on our projects. In addition I will touch upon expectations for future projects and bottlenecks that should to be addressed for the method to become an even more valuable tool in our quest to discover new drugs.

23

Deep Views into Vitrified Cells by Cryo-Scanning Transmission Electron Tomography

S.G. Wolf[1], L. Houben[1], M. Elbaum[2]
[1]Department of Chemical Research Support, [2]Department of Materials & Interfaces, Weizmann Institute of Science, Rehovot, Israel

Technological advances that have brought the "resolution revolution" to macromolecular structure determination by cryoTEM have been providing outstanding results for cellular tomography1, giving first-time glimpses into molecular structure in the cellular context. CryoTEM depends on coherent, elastically scattered electrons to produce phasecontrast, and the sample thickness must be limited approximately to the mean free path (MFP) for inelastically scattered electrons; in practice, <400 nm. Thus, vitrified cells and tissues require thinning by either cryo-microtomy2 or FIB milling3. Scanning TEM (STEM) modality provides a way to probe thicker biological specimens in the vitrified state4, complimenting large-volume methods on plastic-embedded material5-6. Contrast is provided by incoherent detection of electrons, depending on their scattering angle. The natural limit for sample thickness is ~3X larger, at the MFP for elastically scattered electrons. As opposed to cryoTEM, inelastic scattering, while still damaging the sample, provides useful information and improved dose efficiency. CryoSTEM tomography (CSTET)7-9 provides independently and simultaneously collected bright field (BF) and dark field (DF) signals for morphological information (BF), and sensitivity to atomic number (DF), and thus chemical composition. CSTET studies of vitrified prokaryotic and eukaryotic cells up to ~ 1200 nm in thickness10 are presented, highlighting unique functional insights. CSTET fits within the spectrum of methods required for multi-scale visualization of biological systems.

References:

1. Wagner J, et al., FEBS Lett 2017, 591, 2520. 2. Al-Amoudi A, et al.,EMBO J 2004, 23, 3583. 3. Marko M, et al.,J Micr 2006, 222 (1), 42. 4. Wolf SG, et al., Chpt 2 in Cellular Imaging, Electron Tomography and Related Techniques, Hanssen EG, Ed. Springer: 2017 in press. 5. Kremer A, et al.,J Micr 2015, 259 (2), 80. 6. Titze B,Genoud C,Biol Cell 2016, 108 (11), 307. 7. Wolf SG, et al.,J Micr 2015, 260 (2), 227. 8. Kirchenbuechler D, et al.,AIMS Bioph 2015, 2 (3), 259. 9. Wolf SG, et al.,Nat Meth 2014, 11 (4), 423. 10. Rez P, et al.,J Str Biol 2016, 196 (3), 466

24

Keystone Symposia - Poster Abstracts
Cryo-EM from Cells to Molecules: Multi-Scale Visualization of Biological Systems (F1)

Current as of 04/13/2018 09:04

Only the abstract text is displayed here, not the pictures or graphs.

POSTER NUMBER: 1001

Structure of the Nematode Endosymbiont Xenorhabdus nematophilus Toxin Complex A at 3.1 A resolution

Stephen G. Aller[1*], Rachel C. Hart[2], Christopher D. Radka[3], Zhao Wang[4], Joel Sheets[5] and Lawrence J. DeLucas[6]
[1]University of Alabama at Birmingham, [2]Vanderbilt University, [3]St. Jude Children's Research Hospital, 4Baylor College of Medicine, 5Dow Chemical Company and 6AeroSpace Corporation

A 1.7 MDa protein toxin complex is secreted by entomopathogenic *Xenorhabdus nematophilus* and engages cell-surface insect proteins via binding by the A subunit of the toxin complex (TcA). Upon binding, the entire complex is endocytosed into the insect cell where the B/C subunits are cleaved and injected into the cytoplasm to induce insect cell death via an evolutionarily conserved ADP-ribosyltransferase activity. This generalized toxin function would be deadly to most cells and therefore insect cell targeting specificity must be defined exclusively by the function of TcA itself. To better understand the targeting mechanism, we solved the TcA structure by phasing high-resolution x-ray diffraction data using a cryo-EM single-particle reconstruction. During cryo-EM processing, strong 5-fold symmetry was evident in TcA, and the best resolution map (7.0 Å) was achieved by imposing C5 symmetry during 3D particle refinement. X-ray diffraction of SeMet-containing TcA was also achieved, but SAD phasing was unsuccessful most likely from the low-phasing power of selenium and the high number (~305) of expected sites. We therefore took a combined approach and performed "EM-SAD" in which molecular replacement using the cryo-EM map allowed ~42 initial sites to be identified in anomalous difference Fourier electron density maps. In multiple iterative rounds of phasing, increasing SeMet sites could be identified allowing the final experimental electron density map to be phased at 3.5 Å resolution with at least 290 sites. Automated model building was successful using the EM-SAD map, and the final structure was refined to 3.1 Å resolution. The structure reveals that the RBDs form "sockets in grooves", with relatively weak association with the main TcA capsule. A break in the internal 5-fold symmetry (NCS) was observed for at least two of the five receptor binding domains (RBDs) on the surface of TcA that does not appear to influenced by crystal packing. The weak association appears to allow the RBDs to sample slightly different orientations in their respective sockets, which may confer flexibility and allow for some adjustment of their orientations during binding, increasing the binding capacity and efficiency of TcA prior to endocytosis.

1

POSTER NUMBER: 1002

A laser-based Zernike phase plate for cryo-em: recent progress and future prospects

Jeremy J. Axelrod[1], Osip Schwartz[1], Robert M. Glaeser[2,3], Holger Müller[1,2*],
[1]Dept. of Physics, University of California, Berkeley; [2]Lawrence Berkeley National Laboratory, Berkeley, California, USA; [3]Dept. of Molecular and Cell Biology, University of California, Berkeley, California, USA

The realization of a practical Zernike phase plate has long been a goal in electron microscopy. Particularly in cryo-em, such a phase plate would provide increased image contrast thus allowing for the reconstruction of small and/or flexible proteins not otherwise accessible using defocus alone. Ongoing work using the Volta phase plate (VPP) design has shown promising results[1]. However, the VPP suffers from an unstable phase shift and a difficult-to-characterize contrast transfer function (CTF) which complicates its capabilities. Here, we present our recent progress in using a tightly-focused, high-intensity, continuous-wave (cw) laser beam to generate the necessary phase shift to serve as a Zernike phase plate for cryo-em applications[2]. Such a design is inherently immune to charging and phase shift instability, and the CTF can be well-characterized. The primary challenge in implementing this method is the cw high-intensity requirement—approximately 400GW/cm^2 is required to phase shift 300keV electrons by 90°. Such cw intensities have never before been realized. Our approach relies on a near-concentric Fabry-Perot optical resonator to both focus the light and enhance its intensity. Thus far in our optical test setup we have demonstrated intensities sufficient to provide 15° of phase shift for 300keV electrons, or

30° at 80keV. An increase to the target intensity providing 90° at 300keV should be possible with new state-of-the-art resonator mirrors.

[1] Danev and Baumeister, eLife 2016;5:e13046
[2] Schwartz et al. Opt. Express 25, 14453-14462 (2017)

2

POSTER NUMBER: 1003

The structural basis of regulated sphingolipid synthesis

Halil Aydin
University of California, San Francisco, San Francisco, CA 94158

Sphingolipids are constituents of the plasma membrane and bioactive metabolites of profound importance to cellular trafficking, signaling, and inflammation. Sphingolipid synthesis starts in the endoplasmic reticulum (ER) with the condensation of serine and palmitoyl-coenzyme A by serine palmitoyltransferase (SPT) enzyme. SPT activity is negatively regulated by the Orm family of ER membrane proteins, which play an integral role in sphingolipid homeostasis. Elevated expression of Orm-family proteins inhibits the synthesis of sphingolipids and this dysregulation has been implicated in the pathogenesis of asthma and other inflammatory diseases. With the SPT enzyme, Orm-family proteins form an essential enzyme complex, the SPOTS complex, for Serine Palmitoyltransferase, Orm, Tsc3, and Sac1 complex, which catalyzes the rate-limiting step for *de novo* sphingolipid synthesis. However, the molecular architectures of the SPOTS complex and the Orm proteins, as well as the details of their functional roles in sphingolipid metabolism, are entirely unknown. Here, we determined the structure of the SPOTS complex using cryo-electron microscopy and characterized the mechanisms governing sphingolipid homeostasis with biochemistry and cell biology techniques. These results will improve our understanding of sphingolipid homeostasis and shed new light on the pathogenesis of inflammatory diseases.

3

POSTER NUMBER: 1004

Structure of the earliest stable precursor of the ribosome - the small subunit processome

Jonas Barandun[1], Malik Chaker-Margot[1], Mirjam Hunziker[1], Kelly R. Molloy[1], Brian T. Chait[1], Sebastian Klinge[1].
[1]The Rockefeller University, New York

Eukaryotic ribosome biogenesis is a complex and multidimensional process involving all three RNA polymerases, many small nucleolar RNAs and more than 200 assembly factors. The maturation events are initiated in the nucleolus where ribosome biogenesis factors assemble co-transcriptionally on a large precursor RNA. During these early steps, the small subunit (SSU) processome is formed. This giant particle is essential and unique to eukaryotes and catalyzes some of the initial steps in the assembly of the ribosome by a so far unknown mechanism.

Here we present the complete 3.8 Å cryo-EM structure of the SSU processome, a large assembly of more than 4.5 MDa [1]. Electron microscopy in combination with protein-RNA, protein-protein crosslinking and mass-spectrometry allowed us to shed light on the early events in ribosome biogenesis [2-3].

In the SSU processome, U3 snoRNA and 51 assembly factors chaperone a partially processed rRNA bound by 15 ribosomal proteins. The individual domains of the small ribosomal subunit RNA are kept in an open conformation such that enzymes can access RNA regions occluded in the mature ribosome. U3 snoRNA, assisted by several proteins, remodels the central pseudoknot and prevents its premature formation in the particle. An extensive network of peptides and long helices pierce through the particle and contact distant functional sites thereby confirming a hierarchical assembly model. Steric hindrance, molecular mimicry, and exosome recruitment modules suggest that the maturing ribosome has to pass an extensive quality control system for productive ribosome assembly to occur.

[1] Barandun J, Chaker-Margot M, Hunziker M, Molloy KR, Chait BT, Klinge S. *The complete structure of the small-subunit processome.* NSMB. 2017;1-36.
[2] Hunziker M, Barandun J, Petfalski E, Tan D, Delan-Forino C, Molloy KR, et al. *UtpA and UtpB chaperone nascent pre-ribosomal RNA and U3 snoRNA to initiate eukaryotic ribosome assembly.* NCOMMS. 2016;7:12090.
[3] Chaker-Margot M, Barandun J, Hunziker M, Klinge S. *Architecture of the yeast small subunit processome.*

SCIENCE. 2017;355:eaal1880.

This work was supported by an EMBO LTF and a SNSF postdoctoral fellowship.

4

POSTER NUMBER: 1005

Structural characterization of the R-body complex, a macromolecular piston-like nano-machine

Julien R. C. Bergeron[1,3], Jessica K. Polka[2], Charles S. Asbury[1], Pamela A. Silver[2], Justin M. Kollman[1]
[1]Department of Biochemistry, University of Washington, Seattle, USA; [2]Department of Systems Biology, Harvard Medical School, Boston, USA; [3]Current address: Department of Molecular Biology and Biotechnology, The University of Sheffield, Sheffield, UK

The Refractile body (R-body) is a macromolecular assembly first characterized in the bacterium Caedibacter taeniospiralis, a Paramecium endosymbiont. Under physiological conditions, the R-body forms a large, coiled structure in the bacterial cytosol. Upon acidification, the R-body coil undergoes a dramatic architecture rearrangement, converting into an extended form. This change in the architecture of the complex leads to cell lysis, and is exploited by the paramecium host as a mechanism to kill other paramecium cells. Uniquely, the extensive large-scale architectural rearrangement observed for the R-body complex is fully reversible, and does not require any external chemical energy fuel.
Very little is known about the R-body at the molecular level. It has been shown to be encoded by an operon of four genes (rebA-D). However, the role of these in R-body assembly has not been fully elucidated. In particular, there are currently no structural insights into the structure of the Reb proteins, their arrangement within the R-body, and how the change in pH triggers the architecture change.
We have determined that the R-body complex is composed of RebA, RebB and RebD, forming a ribbon-like biomaterial (termed the Reb ribbon). We have characterized this Reb ribbon by cryo-Electron Microscopy and Tomography, demonstrating that it consists of an ordered, repeating lattice, and three separate layers. Finally, we have analyzed its assembly and extension properties, using a combination of fluorescence and Differential Interference Scanning microscopy. Based on these results, we propose a model for the R-body architecture and its pH-mediated rearrangement.

5

POSTER NUMBER: 1006

Heterogeneity and Dynamics in the HIV-1 Envelope Glycan Shield

Zachary T. Berndsen[1], Cesar Lopez[2], Kimmo Rantalainen[1], S. Gnanakaran[2], and Andrew B. Ward[1]
[1]The Scripps Research Institute; [2]Los Alamos National Lab

The HIV-1 Envelope Glycoprotein (Env) is the sole antigen on the virion surface and therefore under strong immune pressure during infection. In response, it has evolved numerous mechanisms to evade the humoral immune system, most important being extensive glycosylation. Env's ~90 N-linked glycans compose >1/2 the mass of the entire protein and extend outward from the protein surface forming a protective barrier known as the "glycan shield." Characterization of broadly neutralizing anti-Env antibodies shows that navigating and accommodating the glycan shield is critical to the development of a protective immune response. However, the flexibility and compositional heterogeneity of long branched glycans hinders crystallization and limits analysis by typical structural biology techniques. Single particle cryo-EM is well suited for studying Env since it is both a tool for solving a high-resolution structure without crystallization, as well as a (quasi)single-molecule technique which analyzes statistical properties of heterogeneous datasets and subsets. Here we present a single-particle cryo-EM analysis of natively glycosylated HIV-1 Env along with complementary all-atom molecular dynamics (MD) simulations. We show how progressive lowpass filtering of high-resolution maps can reveal increasing layers of glycan density including glycan-glycan and glycan-antibody interactions previously hidden below the noise threshold. 3D-variability maps illustrate that the glycan shield is highly dynamic with patches of high and low relative variability usually defined by the local environment. Synthetic cryo-EM data sets created from all-atom MD trajectories and subject to the same analysis reproduce many of the key features seen in both the average density and the 3D-variability maps, providing single molecule level descriptions of the experimental observations. Taken together, these data paint a comprehensive picture of the structure and dynamics of the HIV-1 Env glycan shield and illustrate the utility of converting MD trajectories into synthetic single-particle datasets when studying heterogeneous systems.

This work was supported by the Collaboration for Accelerating Vaccine Design OPP1115782 from the Bill and Melinda Gates Foundation.

POSTER NUMBER: 1007

Cryo EM structure of the activated receptor tyrosine kinase RET signalling complex reveals the importance of its cysteine rich domain

Janna M. Bigalke[1], Shintaro Aibara[2], Robert Roth[1], Alexey Amunts[2], Jenny Sandmark[1]
[1]Discovery Sciences, Innovative Medicines and Early Development Biotech Unit, AstraZeneca, Pepparedsleden 1, Mölndal, 431 83, Sweden; [2]Science for Life Laboratory, Department of Biochemistry and Biophysics, Stockholm University, 17165, Solna, Sweden

The receptor tyrosine kinase RET is activated in the extracellular space by binding to a preformed complex of a GDNF-family-ligand (GFL) and its coreceptor (GFRα). Neurturin (NRTN) is one of four GFLs and predominantly binds to GFRα2. RET activation mediates growth and differentiation signals to various cell types and has been shown *in vitro* and *in vivo* to stimulate survival of neurons. In fact, several GFLs have been studied in clinical trials for treatment of neurological diseases through neuron regeneration. Modulating RET signalling has vast therapeutic potential, as exemplified by recent excitement around the homologous GDF15-GFRαL complex and its implications in treating obesity. Despite their clinical importance, there is no high resolution structural evidence of the GFL-GFRα-RET complex formation and details of receptor activation are unknown. We have determined the first structure of the complete extracellular region of the NRTN-GFRα2-RET complex by cryo EM to a resolution of 5.7Å and thus the first sub-nanometer structure of any GFL-GFRα-RET complex. Although we encountered severe orientation preference of the molecule, we were able to obtain near-atomic resolution data using a recently published tilting protocol. We have identified the residues involved in complex formation and receptor activation, and most importantly we have identified the position of the elusive RET cysteine rich domain (CRD). The structure shows that the CRD is in direct contact with GFRα2 and NRTN, offering an explanation for why RET only binds the preformed complex but not GFRα2 alone. Moreover, our structure agrees well with *in vivo* data from signalling-deficient RET mutants that have been originally discovered in Hirschsprung disease patients. Knowing the details of complex formation makes it possible to manipulate RET activity, which lays the groundwork to discover new therapeutic opportunities to treat a range of neurological diseases such as Alzheimer's, Parkinson's and Huntington's disease.

7

POSTER NUMBER: 1008

Structural analysis of a Parkinson's Disease LRRK2 mutant via cellular cryo-electron tomography and sub-tomogram averaging

Böhning J.[1], Watanabe R.[1], Buschauer R.[1,2], Boassa D.[3], Taylor S.[4], and Villa E.[1]
[1]Department of Molecular Biology, University of California San Diego
[2]Present address: Gene Center, Ludwig-Maximilians University of Munich
[3]The National Center for Microscopy and Imaging Research, University of California San Diego
[4]Department of Chemistry & Biochemistry, Department of Pharmacology, University of California San Diego

With around 6.5 million people affected to date, Parkinson's disease (PD) is the most common neurodegenerative disease after Alzheimer's disease. The cellular phenotype of PD is complex, with implications in protein aggregation, trafficking and turnover having been described, among others. While the majority of cases are sporadic, around 10% of PD patients have a hereditary form of the disease. The most commonly mutated gene in this hereditary form encodes the leucine-rich repeat kinase 2 (LRRK2), a 288 kDa protein with a putative kinase, a GTPase and four protein-protein interaction domains. Since a single mutation in LRRK2 may cause the PD phenotype, understanding LRRK2's function may be key to understanding PD itself. Yet, LRRK2's exact cellular function remains unknown, with implications in several cellular processes having been described. Furthermore, LRRK2 has eluded structural determination, although being a potentially promising drug target.
Recently, several pathological PD mutants of LRRK2 were shown to form filamentous structures around microtubules. As confirmed by functional studies, microtubule-binding is induced by inactivation of LRRK2's kinase and/or GTPase activity by the disease mutation. As disease mutant LRRK2 is trapped on the microtubule, this may suggest that microtubule-bound LRRK2 represents the form of LRRK2 that ultimately causes the PD phenotype. Here, we present a pseudo-atomic model of a PD mutant LRRK2 (I2020T) bound to a microtubule as revealed *by in situ* cryo-electron tomography and subtomogram averaging. Via a combination of high-throughput segmentation, 3D classification and high-resolution refinement, we were able to obtain a high-resolution sub-tomogram average of the microtubule-bound repetitive unit of oligomerized LRRK2. Via integrative modelling, we created a pseudo-atomic model of LRRK2, showing that LRRK2 oligomerizes on microtubule surfaces by forming an inter-connected,

left-handed double-helix, with each LRRK2 monomer forming three dimerization interfaces. Interestingly, in contrast to the canonical microtubules composed of 13 protofilaments found in mammalian cells, LRRK2 (I2020T) exclusively decorates microtubules with 12 protofilaments. These results suggest that mutant LRRK2 might alter the cellular microtubule structure, thus impacting microtubule dynamics and microtubule-associated processes - and potentially leading to the pathological phenotype found in PD patients.

8

POSTER NUMBER: 1033

Structure Based Drug Discovery of Membrane Protein Targets for Medicine that Matters

Mathieu Botte[1], Sandra Markovic-Müller[1], Robert Cheng[1], Nicolas Bocquet[1], Wassim AbdulRahman[1], Sophie Huber[1], Michael Brauchle[1] and Michael Hennig[1]
leadXpro AG
PARK innovAARE, 5234 Villigen, Switzerland
Email: mathieu.botte@leadxpro.com

Today, structure based drug discovery is well implemented in the drug discovery engine of many pharmaceutical companies. Whereas soluble proteins are managed well within the project timelines and portfolio changes in pharmaceutical industry, transmembrane proteins still represent a significant challenge. LeadXpro combines expertise of drug discovery, excellence in membrane protein science, new methods for X-ray data collection at synchrotron and FEL sources like Serial Crystallography and premium access to the synchrotron Swiss Light Source (SLS), the Free Electron Laser (SwissFEL) and single particle cryo-electron microscopy (cryo-EM) at the University of Basel. High quality solubilized and purified membrane proteins can be used to apply biophysical methods including single particle cryo-electron microscopy. Two examples are presented illustrating the progress made to apply biophysical methods for compound screening, hit identification, confirmation and X-ray structural analysis of ligand complexes to inspire chemistry.

9

POSTER NUMBER: 1009

Assembly of Clustered Protocadherin neuronal recognition complexes studied by X-ray crystallography and cryo-electron Tomography

Julia Brasch[1], Kerry M. Goodman[1], Alex J. Noble[2], Venkata P. Dandey[2], Clint S. Potter[2], Bridget Carragher[2], Barry Honig[1,3], Lawrence Shapiro[1]
[1]Zuckerman Mind Brain and Behavior Institute, Columbia University, New York, NY 10027, USA; [2]Simons Electron Microscopy Center, New York Structural Biology Center, The National Resource for Automated Molecular Microscopy, New York, NY 10027, USA; [3]Howard Hughes Medical Institute

Stochastic cell-surface expression of ~60 α-, β-, and γ-clustered protocadherin (Pcdh)isoforms provides vertebrate neurons with single-cell identities that form the basis of neuronal self-recognition. Protocadherins form isoform-specific homophilic *trans* dimers between apposed neuronal membranes, and simultaneously engage in isoform-promiscuous *cis*- dimerization on the same membrane surface. How clustered Pcdhs utilize these two interfaces to assemble between apposed neuronal surfaces remains unknown. Two models for this assembly have been proposed: (1) formation of discrete tetramers, or (2) assembly of polymeric linear zipper-like structures of alternating *cis* and *trans* interactions. Here we combine X-ray crystallography, single-particle cryo-EM and fiducial-less cryo-electron tomography (ET) of full length Pcdh ectodomains to determine the molecular arrangement of the clustered Pcdh self-recognition complex. Despite forming discrete tetramers in solution and single-particle cryo-EM, X-ray crystal structures show clustered Pcdhs to form an extended zipper-like lattice. When clustered Pcdh ectodomains are tethered to liposome surfaces to mimic a native neural membrane environment, they spontaneously assemble at membrane contact sites into highly ordered linear zipper-like assemblies of alternating *cis* and *trans* interactions that are discernable by cryo-ET. The lattice observed in reconstructed tomograms is strikingly consistent with that observed in the crystal structure and mutations targeted to the crystallographically observed interfaces ablate ordered assembly. Our data suggest that formation of linear assemblies by clustered Pcdhs represents the initial step in neuronal self recognition.

10

POSTER NUMBER: 1010

Using evolutionary couplings to help build and refine cryoEM models

Kelly P Brock[1], Chris Sander[1], Debora S Marks[2*]
[1]Dept. of Cell Biology, Harvard Medical School [2]Dept. of Systems Biology, Harvard Medical School
*Corresponding Author

Recent advances in cryoEM have helped deliver unparalleled biological insight; however, the rate at which experimental structures can be solved still lags behind the need to better understand many biomolecules. Our research aims to significantly **accelerate** the accumulation of structural knowledge through computational means, including collaborating with experimental labs to help solve and refine models of protein monomers and complexes. To help bridge the gap between protein sequence and 3D structure, our lab uses a protein's evolutionary history to predict which residues are most likely to physically interact. Identifying potential interacting pairs, or **evolutionary couplings (ECs)**, can be achieved by finding related sequences and applying a global probability model to search for covariation patterns consistent with interacting residues. We have shown that ECs can be sufficient to predict the tertiary fold of a protein [1,2], and thousands of pre-computed models along with our web server can be found at **EVcouplings.org**. Furthermore, we have used ECs to help solve structures using both x-ray crystallography [3] and NMR [4].
Many cryoEM densities are at an intermediate resolution where chain tracing is difficult or impossible. The use of EC distance constraints opens the doors to model building in lower resolution cryoEM densities. We can obtain higher confidence residue placements from low resolution maps by combining EM data with sequence information from ECs. We can also provide models for unknown component structures in cryoEM 3D complex determination, along with predictions about protein-protein contacts. Here, we present the potential for ECs to help build and refine models in conjunction with cryoEM data.

References

1. Marks, DS et al. *Protein 3D Structure Computed from Evolutionary Sequence Variation*. PLoS One 2011.
2. Marks, DS et al. *Protein Structure Prediction from Sequence Variation*. Nature Biotech 2012.
3. Sjodt, M et al. *Structure and function of the bacterial cell wall polymerase RodA*. In review.
4. Tang, Y et al. *Protein structure determination by combining sparse NMR data with evolutionary couplings*. Nature Methods 2015.

11

POSTER NUMBER: 1011

Role of the dileucine motif in Nef-induced trimerization of the Arf1:AP-1 clathrin adaptor complex

Cosmo Z. Buffalo[1], Kyle L. Morris[1], Xuefeng Ren[1], and James H. Hurley[1]
[1]Department of Molecular and Cell Biology and California Institute for Quantitative Biosciences, University of California, Berkeley, Berkeley, CA 94720, USA

HIV Nef downregulates cell surface host proteins, including SERINCs, CD4, and MHC-I, by redirecting them into clathrin coated vesicles (CCVs) for lysosomal degradation. Nef achieves this by hijacking the AP-1 and AP-2 clathrin adaptors. HIV-1 Nef cooperates with the host GTPase Arf1 to induce oligomerization and activation of AP-1, assembling as trimers in vitro. AP complexes' normal function is to sort two main classesof cargo proteins into CCVs. These cargo classes aredefined by the presence of a tyrosine-based YXXΦ motif or an acidic-dileucine motif DE]XXXL[LI]. Nef-induced downregulation of MHC-I involves the hijacking of both the the tyrosine and dileucine cargo binding sites of AP-1. Arf1 promotes dimerization of AP-1, while in the presence of Arf1 and Nef together, AP-1 trimers predominate. The combination of dimer and trimer interfaces lead to formation of a hexagonal lattice. In order to determine the contribution of the Nef dileucine motif to trimerization, a cryo-EM analysis of the AP-1:Arf1:NefLLAA complex was carried out. From this analysis, we conclude that the dileucine motif of Nef is important for promoting the trimerization of the Arf1:AP-1 clathrin adaptor complex and subsequent clathrin cage assembly.

12

POSTER NUMBER: 1012

Human IMPDH1 retinal degeneration mutations promote polymer assembly

Anika L. Burrell[1], Matthew C. Johnson[1], Justin M. Kollman[1]
[1]University of Washington

Dynamic assembly of metabolic enzymes is a widespread phenomenon in the cell and provides a mechanism for the physical organization and regulation of cellular metabolism. Large-scale filament assemblies exist in both purine and pyrimidine biosynthesis - pathways that require tight regulatory control to maintain nucleotide homeostasis. In the purine biosynthesis pathway in humans, the enzyme IMP dehydrogenase (IMPDH), which catalyzes the first committed step in GTP synthesis, forms dynamic, microns-long filamentous structures in cells. We seek to characterize IMPDH filaments of both human IMPDH isoforms (IMPDH1 and IMPDH2) to better understand the role IMPDH plays in regulating purine biosynthesis. Furthermore, single mutations in IMPDH1 have been linked to retinal degeneration, underscoring the essential role of nucleotide homeostasis in preventing disease.

Using cryo-electron microscopy, we have determined that the IMPDH octamer building block can adopt an extended (active) or compressed (inactive) state in the filament. We have solved a sub-nanometer structure of the compressed IMPDH2 octamer and a 3.6Å structure of the IMPDH2 filament in an extended state. Current studies are aimed at extending this work to the other IMPDH isoform, IMPDH1. Point mutations in IMPDH1 have been linked to retinal degeneration in humans and our data reveal that many of these mutations cause IMPDH1 filaments to associate into large bundles with varied architectures. These mutations stabilize the filament under conditions that differ from wild type, which could result in a disruption of the normal regulatory role of IMPDH filaments in the cell. The link between these mutations, polymer stabilization, and retinal degeneration demonstrate the functional importance of polymerization and the necessity to correctly regulate metabolic enzymes involved in nucleotide biosynthesis.

13

POSTER NUMBER: 1013

Cryo-EM Structure of Orco: The Molecular Basis for Insect Olfaction

Joel A. Butterwick, Josefina del Mármol*, Kelly H. Kim*, Martha A. Kahlson, Jackson Rogow, Thomas Walz and Vanessa Ruta†
The Rockefeller University, New York, NY
†corresponding author
*equal contributions

Insects are the most diverse group of organisms on Earth, representing approximately half of all identified eukaryotic species on the planet. The success of insects reflects their remarkable capacity to adapt to a wide variety of ecological niches. The rapid evolution of insect olfactory receptors is thought to contribute to this adaptation, endowing each insect species with the ability to selectively detect volatile chemicals associated with their specialized habitat and lifestyle. Insects have evolved a unique family of olfactory receptors, which are heteromeric odorant-gated ion channels comprised of two subunits—a highly divergent odorant receptor (OR) subunit that confers odorant specificity, and a conserved common co-receptor (Orco) subunit. Orco is present in the olfactory sensory neurons of basal insects that lack ORs and likely represents the ancestral form of the olfactory receptor complex. Indeed, Orco subunits themselves form homomeric cation channels that are responsive to synthetic agonists. As insect odorant receptors lack homology with any other protein family, many of their most elementary functional and structural characteristics have remained elusive, such as their stoichiometric organization and the location of their ion conduction pore. Here we used cryo-electron microscopy to determine the first structure of an Orco homomer, in complex with mouse monoclonal antibody fragments, at a resolution of 3.5 Å. Orco subunits are symmetrically arranged around a central aqueous pore that electrophysiological measurements confirm forms the ion conduction pathway. The structure of Orco also provides insight into how the unique architecture of this receptor family may accommodate the substantial sequence variation of the ORs and facilitate the rapid evolution of odor tuning. Orco thus serves as the founding member of perhaps the largest and most diverse family of ion channels in nature.

14

POSTER NUMBER: 1014

Design and Construction of Membrane Surface Complexes for Magnetic Resonance and CryoEM Studies

Jess Li[1], Yifei Li[1], Cameron McGlone[1], Olivier Soubias[1], Paul A. Randazzo[2], R. Andrew Byrd[1]
[1]Structural Biophysics Laboratory, NCI; [2]Laboratory of Cellular and Molecular Biology, NCI

Many biological processes operate at the membrane surface and involve either integral membrane receptors or complexes that assemble at the membrane surface. We are interested in a multi-disciplinary approach to studying complexes involving ADP-ribosylation-factors (Arfs) and Arf-specific GTPase activating proteins (ArfGAPS), which can regulate Arfs by stimulating their slow intrinsic GTP hydrolysis. These complexes assemble using at least two association events: (i) myristoylation of Arf leading to insertion in the membrane and localization, and (ii) binding of a PH domain within the ArfGap to the PIP2-headgroup of lipids within the membrane (eg. Jian,..., Byrd, Randazzo, *Structure* 2015). To facilitate structural and biophysical studies of these system, we have employed nanodiscs of defined composition, combined with appropriate myristoylation of Arf, and sophisticated isotopic labeling for NMR and/or pulsed EPR DEER measurements. In forming these complexes, control and knowledge of the composition is critical to high quality data. The same will be true for applying cryoEM to these systems. We have utilized NMR and SAXS methods to characterize and quantitate the composition and uniformity of these systems. These analyses suggest that the utilization of more homogeneous structures is important for cryoEM studies. In this regard, we are exploring the use of circularized membrane scaffolding protein (Nasr, Hagen, Wagner, et al. *Nature Methods* 2017) to construct the nanodiscs. These assemblies provide the ability to obtain solution NMR and DEER EPR data, and they will be used to screen for adequate sampling conditions to yield medium to high resolution cryoEM images for reconstruction.

15

POSTER NUMBER: 1015

Structural Snapshots of a Dynamic Process in the Regulation of P-Rex, a Metastatic Factor

Jennifer N. Cash[1], Ellen M. Davis[1], Michael A. Cianfrocco[1], and John J.G. Tesmer[2]
[1]Life Sciences Institute and Department of Biological Chemistry, University of Michigan, Ann Arbor, MI 48109; [2]Department of Biological Sciences, Purdue University, West Lafayette, IN 47907

Phosphatidylinositol 3,4,5-trisphosphate (PIP_3)-dependent Rac exchanger (P-Rex) is a Rho guanine-nucleotide exchange factor that regulates cell motility and is strongly associated with cancer metastasis. P-Rex is synergistically recruited to the cell membrane and activated by PIP_3 and heterotrimeric G protein βγ subunits, positioning the enzyme downstream of multiple classes of cell surface receptors. P-Rex has thus become an attractive therapeutic target for the suppression of cancer metastasis. However, development of selective inhibitors against P-Rex is hindered by the fact that its regulatory mechanisms are poorly understood. To address the role of conformational dynamics of P-Rex, we have taken an integrative structural biology approach using X-ray crystallography, small-angle X-ray scattering (SAXS), hydrogen/deuterium exchange mass spectrometry (HDX-MS), and cryo-electron microscopy (EM) to define the molecular basis for the regulation of P-Rex, with the aim of identifying its important regulatory surfaces and mechanisms of activation. We have determined crystal structures of the P-Rex1 tandem Dbl homology/pleckstrin homology (PH) domain catalytic core and shown that the PH domain is necessary and sufficient for PIP_3-dependent activation. Our data is consistent with PIP_3 activating P-Rex1 through an allosteric mechanism as opposed to simple recruitment to the cell membrane. Using these structures and data derived from HDX-MS and SAXS experiments, we have developed a model for how P-Rex is regulated. To test this model, we are analyzing cryo-EM data of P-Rex1 with the goal of providing high-resolution information for the full-length protein. By investigating P-Rex through a variety of complementary structural approaches, we hope to guide the future development of therapeutic molecules.

Work supported by an American Cancer Society - Michigan Cancer Research Fund Postdoctoral Fellowship (PF-14-224-01-DMC) to J.C., an American Heart Association Undergraduate Student Research Program Award (14UFEL20510027) to E.D. and NIH National Heart, Lung, and Blood Institute Grants HL071818 and HL122416 to J.T.

16

POSTER NUMBER: 1037

Dynamo: flexible software tools for cryoelectron tomography data.

Daniel Castaño Díez[1*], Paula Navarro, Steffano Scramuzza[2], Henning Stahlberg[2]
[1]BioEM-Lab, Biozentrum, University of Basel; [2]C-CINA, Biozentrum, University of Basel

While Cryo Electron Tomography offers a unique potential for direct visualization of the three dimensional structure of macromolecular compounds in their cellular environment, the severely noisy nature of the low dose images delivered by the microscope makes this imaging technique heavily dependent on a suitable computational treatment of the raw data. In particular, structure determination through Subtomogram Averaging requires the localization of large number of noisy copies of the compound of interest and extraction of the common signal through classification and alignment.

This presentation focuses on several data processing challenges appearing in different stages of this technique, discussed in the context of the unified processing framework provided by the Dynamo software package. The new Catalogue module covers the logistical tasks attached to the organization of large datasets, providing specific tools for visualization and geometrical modeling of tomogram content by manual or semiautomated identification and annotation of structures as membranes, filaments or lattice-like arrangements of particles.

Further modules address the analysis of the identified subtomograms, guiding the users through the use of standardized analysis steps or rather supporting the design of new, case specific approaches. The software can harness High Performance Computing environments like CPU or GPU clusters, in order to cope with the numerically demanding nature of the technique, an aspect that will be showcased by different case studies.

17

POSTER NUMBER: 1034

The molecular basis for UV-damage recognition in chromatin

Simone Cavadini[1], Syota Matsumoto[1], Richard D. Bunker[1], Shigenori Iwai[2], Hitoshi Kurumizaka[3], Kaoru Sugasawa[4], Nicolas H. Thomä[1]
[1]Friedrich Miescher Institute for Biomedical Research, Maulbeerstrasse 66, 4058 Basel, Switzerland
[2]Department of Chemistry, Graduate School of Engineering Science, Osaka University, 1-3 Machikaneyama, Toyonaka, Osaka 560-8531, Japan
[3]Laboratory of Structural Biology, Graduate School of Advanced Science and Engineering, Waseda University, 2-2 Wakamatsu-cho, Shinjuku-ku, Tokyo 162-8480, Japan
[4]Biosignal Research Center, Kobe University, 1-1 Rokkodai-cho, Nada-ku, Kobe-shi, Hyogo 657-8501, Japan

DNA in each cell is constantly exposed to several environmental and endogenous threats. This leads to a variety of DNA mutations that occur spontaneously (i.e., replication errors), after exposure to reactive chemical intermediates (i.e., reactive metabolites), or following exposure to UV-light. The global genome repair (GGR) pathway surveils genomic DNA for UV-light induced pyrimidine dimers, and targets these for repair by the nucleotide excision repair (NER) pathway (Ref. 1). Cyclo-butane pyrimidine dimers (CPDs) and pyrimidine-pyrimidone 6-4 photoproducts (6-4PP) are found throughout the genome, including regions of chromatin, and are detected by the UV-damaged DNA binding proteins (UV-DDB complex), comprised of DDB1 and DDB2. How DNA lesions are sensed when embedded within nucleosomes, with histones and the two DNA gyres potentially restricting accessibility to the damage, is unclear at present.
We set out to investigate structurally and biochemically how the UV-DDB complex binds damaged nucleosomes around the sites where the minor groove points outward, a location where CPDs are frequently found following UV-irradiation in model systems in vivo, and cell lines (Ref. 2). We present atomic resolution cryo-EM structures of UV-DDB in complex with nucleosomes containing 6-4PP or abasic sites that detail UV-damage recognition at different sites. We find that the recognition of more distal lesions from the most populated CPDs site is not possible on strongly positioned nucleosomes. We describe how UV-DDB recognizes UV-induced damage in chromatin, and the emerging role of nucleosome positioning strength in DNA damage recognition.
Our study provides the molecular basis for UV-damage detection by UV-DDB in chromatin, and offers a conceptual framework for DNA sequence/damage detection by diverse DNA binding proteins, including transcription factors.

References

1. 1) Sugasawa, K., Mutat. Res. 685, 29-37 (2010).
2. 2) Pfeifer P. G., Photochemistry and Photobiology, 270-283(1997).

18

POSTER NUMBER: 1016

Structure analyses of vaccinia viral A26 protein that is critical for mature virus endocytosis

Hung-Wei Chang[1], Huei-Yin Cheng[1], Kathleen Joyce D. Carillo[2,3,4] and Der-Lii M. Tzou [2,3,5] and Wen Chang[1*]
[1]Institute of Molecular Biology, Academia Sinica, [2] Sustainable Chemical Science and Technology Program, Taiwan International Graduate Program, Academia Sinica, [3]Institute of Chemistry, Academia Sinica, Taipei, Taiwan, ROC [4] Department of Applied Chemistry, National Chiao Tung University, Hsinchu, [5] Department of Applied Chemistry, National Chia-Yi University, Chia-Yi, Taiwan, ROC

Vaccinia mature virus (MV) enters cells through either endocytosis or plasma membrane fusion, depending on cell types and virus strains. Our previous studies showed that WR strain of vaccinia mature virus enters HeLa cells through fluid phase endocytosis. Interestingly, when we deleted an ORF encoding a MV membrane A26 protein from vaccinia viral genome the resulting deletion virus, WRΔA26, infects HeLa cells through plasma membrane fusion. We thus hypothesized that A26 protein may act as an acid-sensitive membrane fusion suppressor protein that controls virus entry pathway specificity. Here we aim to dissect the structure domain of A26 protein in order to understand how it functions as an acid-sensitive regulator of membrane fusion. We thus generated a series of recombinant vaccinia viruses that contains various A26 mutant proteins, purified these MV particles and investigated how these mutant viruses enter cells. We also generated and purified different segments of recombinant A26 protein and investigate protein conformation *in vitro*. Together, our data will provide important genetic and biochemical evidence for understanding A26 protein function during virus entry.

19

POSTER NUMBER: 1017

Single particle Cryo-EM studies of mitochondrial inner-membrane fusion

Luke H. Chao[1,2]
[1]Department of Molecular Biology, Massachusetts General Hospital; [2]Department of Genetics, Harvard Medical School

Mitochondrial membrane dynamics are intimately tied to the organelle's diverse functions. Modulating the ratio of mitochondrial fission and fusion is a regulatory node that directly connects metabolic state with organelle morphology. We present single particle Cryo-EM reconstructions of the inner mitochondrial membrane fusogen OPA1 that reveal multiple conformational states. We discuss relationships between these states with the protein's fusogenic conformational rearrangement as well as potential roles for these conformers in the maintenance and rearrangement of cristae. These studies provide a framework for relating conformational change in mitochondrial fusion proteins with their membrane-associated activities.

20

POSTER NUMBER: 1018

Super-resolution Single-Particle Electron Microscopy

James Chen
Department of Biochemistry and Molecular Biology, OHSU

The Nyquist Frequency represents the information cut-off in discrete signal sampling. In single-particle EM, the Nyquist Frequency dictates the highest resolution achievable in 3D reconstruction. However, this information upper-limit stands only for a single image. When the same object is recorded into a movie stack with random subpixel frame translations, the image ensemble actually contains information beyond the Nyquist Frequency. In this work, an algorithm is proposed and implemented to retrieve such information from 2D image series. Its application in single-particle cryo-EM is demonstrated on an experimental movie dataset, in which image stacks recorded at 2.2 Å/pixel have enabled the 3D reconstruction reaching 3.3 Å resolution.

21

POSTER NUMBER: 1035

CryoEM structure of Mayaro virus: a re-emerging New World alphavirus

David Chmielewski[1], Jason T. Kaelber[2], Jonathan A. Auguste[3], Wah Chiu[1]
[1] Biophysics Program, Stanford University, Palo Alto, CA 94305 USA
[2] Center for Integrative Proteomics, Rutgers University, New Brunswick, NJ 08901
[3] College of Agriculture and Life Sciences, Virginia Tech University, Blacksburg, VA 24061

Mayaro virus (MAYV) is a mosquito-borne pathogen that causes acute febrile illness, rash, nausea and severe, persistent arthralgia in humans. The virus is endemic to jungle-containing regions of Central and South America, and a recent case in Haiti suggests active circulation in the Caribbean. MAYV is a member of the *Alphavirus* genus of family *Togaviridae*, and grouped among the arthralgic Old-World alphaviruses along with Chikungunya (CHIKV) and Semliki Forest Virus (SFV) among others. A separate grouping of alphaviruses termed "New-World" based on geographic distribution is characterized by lethal encephalitis phenotype. Alphaviruses are icosahedral, enveloped, single-strand RNA viruses ~70 nm in diameter. All alphavirus particles consist of an inner protein nucleocapsid core that binds the RNA genome, a host-derived membrane bilayer and an outer glycoprotein shell formed by trimers of E1-E2 heterodimers in t=4 icosahedral symmetry. Using cryo-electron microscopy, we determined the structure of wild-type MAYV to a resolution of 4.0 Å. This density map provides the most detailed structure of an alphavirus virion to date, allowing all-atom modeling and understanding of virus assembly.
Since the first structures of *Alphavirus* virions, an orientation difference has been observed between the nucleocapsids of encephalitic New-World and arthralgic Old-World viruses. Comparison of all-atom models of MAYV (Old-World) and Venezuelan equine encephalitis virus (VEEV) (New-World) revealed key structural differences in flexible domains in and around the lipid envelope that are resolved in the near-atomic cryoEM maps. E1-E2 glycoprotein interactions in the protein endodomain, are shown critical to providing the proper spike-capsid

interactions for the formation of the double-shelled icosahedral virion.

22

POSTER NUMBER: 1019

Deciphering the structure of a novel class of cholesterol-dependent cytolysin like proteins

Michelle P. Christie[1], Sara L. Lawrence[1], Bronte Johnstone[1], Craig J. Morton[1], Rodney K. Tweten[2] and Michael W. Parker[1,3]
[1]St Vincent's Institute for Medical Research, Fitzroy, VIC 3065, Australia [2]The University of Oklahoma Health Sciences Center, Oklahoma City OK 73104, USA, [3]Bio21 Molecular Science and Biotechnology Institute, Parkville, VIC 3010, Australia

The cholesterol-dependant cytolysins (CDC) are a family of protein toxins found largely in Gram-positive bacteria. CDCs play an important role in bacterial pathogenesis and function by introducing large ring-like pores into cholesterol containing eukaryotic membranes. CDCs are secreted as water-soluble monomers that undergo large conformational changes to oligomerise and insert into target membranes. Membrane insertion disrupts intact cells causing cell death [1].
CDCs derived from Gram-positive bacteria contain four domains [2,3], of which the membrane-binding region in domain 4 and a tyrosine and glycine rich sequence found in the membrane-spanning region in domain 3 are highly conserved [4]. Recent in silico studies of protein sequences identified a hitherto uncharacterised family of proteins in Gram-negative bacteria that contained the conserved tyrosine and glycine rich sequence. However, significant differences exist in the membrane-binding domain 4 region.
The aim of our work is to characterise the structure and function of this new family of CDC like proteins. To do so we recombinantly expressed and purified CDC like proteins from three Gram-negative species and tested their ability to interact with lipid membranes. Our experiments showed that similar to the Gram-positive CDCs these proteins could assemble on membranes to form large oligomers. However, the conditions under which they oligomerised were different. We have obtained protein crystals of the monomeric form for some of the proteins. Preliminary electron microscopy experiments showed that the CDC like proteins form large ring like pores similar to their Gram-positive counterparts, although their size and morphology are different.
References
[1] Eileen M. et al. (2012) Biochim Biophys Acta (Biomembranes) 1818 (4) 1028
[2] Lawrence S. L., et al. (2015) Sci Rep 5:14352 doi: 10.1038/srep14352
[3] Feil S.C., et al. (2014) J Mol Biol. 426 (4):785
[4] Farrand A. J. et al., (2015) J. Biol. Chem. 290 (29) 17733

23

POSTER NUMBER: 1020

Cryo-EM studies of the human sterol transporter G5G8 in nanodiscs

Lindsay D. Clark[1], Jin Wang[2], Xiaochen Bai[1], Xiao-Song Xie[2], Jonathan C. Cohen[2], Helen H. Hobbs[2,3], Daniel M. Rosenbaum[1]
[1]Department of Biophysics, University of Texas Southwestern Medical Center, Dallas, Texas 75390; [2]Eugene McDermott Center for Human Growth and Development, University of Texas Southwestern Medical Center, Dallas, Texas 75390; [3]Howard Hughes Medical Institute, University of Texas Southwestern Medical Center, Dallas, Texas, 75390

Cholesterol is a fundamental component of mammalian cell membranes, and maintenance of appropriate levels of cholesterol is crucial to membrane integrity and cell viability. The heterodimeric ABC transporter G5G8 plays a key role in cholesterol homeostasis by promoting the efflux of cholesterol and plant sterols into bile (1). Deleterious mutations in G5 or G8 cause the rare genetic disorder sitosterolemia, which leads to hypercholesterolemia and premature atherosclerosis (2). Our lab previously solved the crystal structure of human G5G8 in DMPC/cholesterol bilayers (3), providing the first structural insights into how this transporter functions. The structure revealed G5G8 in a nucleotide-free state, with additional density corresponding to bound cholesterol on the membrane-facing side of the TM bundle. Efforts to crystallize the transporter in different states were unsuccessful, prompting us to utilize cryo-EM to fully characterize the effects of nucleotide and sterol binding on G5G8. To improve contrast and increase total molecular weight, we generated and characterized conformationally-selective antibodies to G5G8 reconstituted in lipid nanodiscs. We have preliminary 3D reconstructions showing well-defined density for G5G8 with a Fab unambiguously bound to one of the nucleotide-binding domains. Efforts are currently underway to solve cryo-EM structures of G5G8 in nanodiscs containing cholesterol, either in apo or ATP-trapped states, which will provide a full mechanistic understanding of how G5G8 mediates dietary sterol balance. In the future, we can extend our expertise to other integral membrane proteins involved in sterol homeostasis to understand the effect of lipids and sterols on these important regulatory proteins.

References
1. Wang, J. *et al. J. Lipid Res.* (2014) vol. 56 (2) pp. 319-330
2. Berge, K. E. *et al. Science* (2000) vol. 290 (5497) pp. 1771-5
3. Lee, J-Y. *et al. Nature* (2016) vol. 533 (7604) pp. 561-4

24

POSTER NUMBER: 1021

Silicon Nitride as a Platform for Cryo-Electron Tomography Based Structural Cell Biology

Matthew Croxford*, Vinson Lam*, Reika Castillon, Elizabeth Villa
University of California San Diego, San Diego, CA, USA

Cell motility is a structurally dynamic process that involves cell-wide rearrangements in cytoskeletal structure and nuclear signaling to redirect cellular activity in response to chemical, thermal, electrical, and mechanical stimuli. In order to study the cytoskeleton and nuclear periphery at an atomic level, we are developing a new set of tools and techniques to prepare *in vivo* vitrified cells to capture highly dynamic processes.
Cryo-Electron Tomography (CET) has traditionally required cells be cultured or deposited on grids comprised of a thin layer of carbon supported by metal grids, followed by rapid freezing to preserve cells and molecules in a native intact state. However, there are a number of reasons that the carbon film substrate is not ideal: (1) they are atypical, physiologically irrelevant cell culture surfaces, (2) they are fragile, and (3) the soft surface favors cell adhesion directly on top of the metal crossbars, which are opaque to TEM and therefore unusable for CET. As an alternative substrate, we propose to use EM grids with a solid, continuous silicon nitride (SiN) surface on which cells can be directly cultured. These properties make SiN a more biologically relevant surfaces compared to carbon grids, but existing techniques need to be adapted to make these viable for CET. To this end we have explored surface treatments to culture cells on SiN grids and are developing methods for manipulating cells to study dynamic cellular processes.
Because SiN grids are not perforated like traditional foil grids, specimens cannot be prepared through blotting. To address these difficulties, we have designed a prototype laminar air flow plunger that directs a blade of air to sweep off excess liquid immediately before vitrification. With this device, we are able to create frozen, hydrated specimens that preserve the dynamic features of cells cultured directly on SiN grids. Using such samples, we plan to study cell motility at an atomic level by using FIB milling to produce cell sections only ~150 nm thick and CET to acquire high-resolution tomograms and to open windows into the cell. We are excited to continue development of SiN grids as a generalized cryo-EM substrate to study a wide variety of highly dynamic cell processes in a near-native state.

25

POSTER NUMBER: 1023

Use of Amphipathic Polymers in Preserving Native Lipids for Membrane Proteins in Cryo-EM

Simon P. Davies[1], Jonathan D. Lippiat[1] and Stephen P. Muench[1]
[1]School of Biomedical Sciences, University of Leeds, UK

Membrane protein structural biology has been revolutionised by the advancements in electron microscopy, however significant challenges remain, not least extraction of membrane proteins from lipid bilayers. Moreover, it is essential that the proteins be recovered in an active, folded state to allow the structure to fully reflect the native state. Historically, detergent-based extraction techniques have been used for membrane protein preparation, however these are often poor at replicating the natural environment of the membrane lipid bilayer and can strip away closely associated lipids.
In recent years, amphipathic polymer molecules known as styrene maleic acid (SMA) and diisobutylene maleic acid (DIBMA) have emerged as promising candidates for the extraction of membrane proteins surrounded by their native lipids. SMA polymers insert into lipid bilayer and extract discs of lipid out of the membrane, so that membrane proteins can be recovered surrounded by natively associated lipids, which are often essential for protein structure and function. Exploring the capabilities of SMA and DIBMA may therefore provide fruitful avenues for the preparation of membrane proteins that better reflect the native conformation.

We have tested SMA and DIBMA against a series of human potassium channel constructs expressed in HEK293S GnTI⁻ cells, with DDM employed in parallel to assess the effectiveness of protein extraction. The identity and quality of the extracted proteins has been assessed using circular dichroism and mass spectroscopy. This was followed by structural analysis using negative stain and cryo-EM.
Using SMA and DIBMA, we have been able to successfully extract and purify a variety of human potassium channel proteins with results comparable to preparation using DDM. Low resolution negative stain reconstructions have been generated. Efforts are currently underway to optimise analysis using cryo-EM, with the goal of generating higher resolution structures in the near future.

26

POSTER NUMBER: 1025

Architectures of lipid transport systems for the bacterial outer membrane

Damian C. Ekiert[1], Gira Bhabha[1]
[1]The Skirball Institute, NYU School of Medicine

The bacterial outer membrane is a lipid bilayer that plays a key role in resistance to antibiotics, detergents, and other external stresses. Despite decades of research on the bacterial envelope, it is unknown how phospholipids are trafficked between the bacterial inner and outer membranes, through the intervening hydrophilic space of the periplasm. We recently discovered that members of the mammalian cell entry (MCE) protein family form structurally diverse hexameric rings and barrels, and that some of these proteins may form "bridges" or "pipes" between the inner and outer membrane to facilitate lipid transport (Ekiert, *et al.* Cell 2017). We are now working to understand how MCE proteins help to create a periplasm-spanning transport system that links the inner and outer membranes. Using cryo-EM and X-ray crystallography, along with complementary genetic and biochemical approaches, we are unraveling how the components of the *E. coli* Mla, Pqi, and Yeb systems assemble into larger inner membrane, outer membrane, and even transenvelope complexes. This work will advance our understanding of a fundamental yet poorly understood aspect of bacterial cell biology, and may open up avenues to the development of new antibiotics that target the essential process of outer membrane biogenesis. In addition, the presence of MCE proteins in some double-membraned organelles, such as chloroplasts, suggests that understanding *E. coli* MCE systems will also have direct implications for lipid trafficking in other bacterial-derived organelles.

27

POSTER NUMBER: 1026

Ribosome-Associated Vesicles: a dynamic vesicular endoplasmic reticulum in secretory cells

Stephen D. Carter[1,§], Cheri M. Hampton[2,§], Robert Langlois[3,§], Roberto Melero[4], Zachary J. Farino[5], Wen Li[3], Ngoc Han Tran[6], Robert A. Grassucci[3,7], Stephanie Siegmund[8], Joshua Pemberton[9], William J. Rice[10], Christoph Wigge[10], Peter Walter[6], Tamas Balla[9], Jose Maria Carazo[4], Grant J. Jensen[11*], Joachim Frank[2,3,7*], Zachary Freyberg[5*]
[1]Division of Biology and Biological Engineering, California Institute of Technology
[2]Howard Hughes Medical Institute (HHMI); Department of Biochemistry and Molecular Biophysics, Columbia University
[3]Department of Biochemistry and Molecular Biophysics, Columbia University
[4]Biocomputing Unit, Centro Nacional de Biotecnología - CSIC
[5]Department of Psychiatry, University of Pittsburgh
[7]Department of Biological Sciences, Columbia University
[6]HHMI; Department of Biochemistry and Biophysics, University of California, San Francisco
[8]Department of Neurology, Columbia University
[9]Section on Molecular Signal Transduction, Program for Developmental Neuroscience, National Institute of Child Health and Human Development, National Institutes of Health
[10]New York Structural Biology Center
[11]HHMI; Division of Biology and Biological Engineering, California Institute of Technology
[§]These authors contributed equally to this work
[*]Corresponding authors

The endoplasmic reticulum (ER) communicates with other organelles, including mitochondria, at discrete contact sites along an extensive network of membranes. However, the structural basis for forming and maintaining these contacts is poorly characterized. Here we combine confocal microscopy in living cells with *in situ* cryo-electron microscopy, tomography and correlative light and electron microscopy (CLEM) to directly visualize the interactions between ER and mitochondria at high resolution under near-native conditions. Using these imaging approaches, we identify a novel, mobile form of ER, Ribosome-Associated Vesicles (RAVs), found primarily in the cell periphery. RAVs exist as distinct structures separate from the intact ER reticular architecture and form dynamic associations with mitochondria via direct intermembrane contacts. We also show that RAVs are conserved across multiple cell types and species. Ultimately, our analyses further establish the diversity of ER subtypes within cells and provide the basis for a better understanding of how ER communicates with other organelles.

28

POSTER NUMBER: 1027

Lipid Bilayer Structure in Native Cell Membrane Nanoparticles of Multidrug Exporter AcrB

Weihua Qiu[1,2,*], Ziao Fu[3,*], Guoyan Xu[1], Robert A. Grassucci[4,5], Wayne A. Hendrickson[4,5,6,7,8], Yan Zhang[1], Joachim Frank[4,5,9], Youzhong Guo[1,2]
[1]Department of Medicinal Chemistry, Virginia Commonwealth University, Richmond, VA 23298, USA; [2]Institute for Structural Biology, Drug Discovery and Development, Virginia Commonwealth University, Richmond, VA 23219, USA; [3]Integrated Program in Cellular, Molecular, and Biomedical Studies, Columbia University, New York, NY 10032, USA; [4]Department of Biochemistry and Molecular Biophysics, Columbia University, New York, NY 10032, USA; [5]Howard Hughes Medical Institute, Columbia University, New York, NY 10032, USA; [6]The Center on Membrane Protein Production and Analysis (COMPAÅ), New York Structural Biology Center, 89 Convent Avenue, New York, NY 10027, USA; [7]New York Structural Biology Center, Synchrotron Beamlines, Brookhaven National Laboratory, Upton, NY 11973, USA; [8]Department of Physiology and Cellular Biophysics, Columbia University, New York, NY 10032, USA; [9]Department of Biological Sciences, Columbia University, New York, NY 10027, USA

Structural information for cell membrane and protein lipid interaction is in high-demand for understanding the function of many membrane proteins; however, this information is still rare because currently there is no efficient approach to obtain this information. We recently developed a novel native cell membrane nanoparticles system. With this system, we determined the structure of native cell membrane lipid bilayer that associated with a membrane protein, multidrug exporter AcrB, using single-particle cryo-EM. Our cell membrane lipid bilayer model consists of 12 lipid molecules in the inner leaflet and 12 lipid molecules within the outer leaflet of the lipid bilayer. The lipid bilayer interacts with protein specifically through both phosphate head and hydrocarbon tails. The hydrocarbon tails within the inner leaflet of the lipid bilayer are in a well-organized hexagonal pattern. Our results provided valuable structural information for the interaction between AcrB and cell membrane. We demonstrated the great potential of using native cell membrane nanoparticles system in membrane protein structural biology.

29

POSTER NUMBER: 1028

Exploration of eukaryotic nuclei at molecular resolution

Shujun Cai[1], Chen Chen[1], Yajiao Song[1], Hong Hwa Lim[3,4], Jian Shi[1], Sachiko Tamura[5], Kazuhiro Maeshima[5], Uttam Surana[2,3,4], Lu Gan[1*]
[1]Centre for BioImaging Sciences and Depts. of Biol. Sciences & [2]Pharmacology, National University of Singapore; [3]Institute of Molecular and Cell Biology, A*STAR, and [4]Bioprocessing Technology Institute, Singapore; [5]National Institute of Genetics and Sokendai, Japan
*Corresponding Author

Chromosome organization at the nucleosome level has long been proposed to influence nuclear activities like transcription, replication, recombination, and chromosome segregation. We have taken advantage of advances in electron cryotomography (cryo-ET) to explore the *in situ* organization of chromatin in single-celled eukaryotes such as picoplankton and yeasts. In three organisms, we did not find any evidence of 30-nm fibers, though we cannot rule out the existence of very short nucleosome chains that fold this way. Furthermore, we did not find any structures that resemble the topologically associated domains depicted in the literature. Instead of packing as large globular structures, chromatin is organized as much-smaller clusters containing fewer than 100 nucleosomes. In mitotic fission yeast, the nucleosome clusters become slightly larger, but the chromosomes do not condense into monolithic structures. Mitotic chromatin is porous, with large ribosome-sized macromolecular complexes and small nucleosome-free pockets interspersed within. Therefore, chromatin is permissive to high levels of transcription throughout the cell cycle and regulate its expression programs without large-scale changes in nucleosome packing.

30

POSTER NUMBER: 1029

The cryo-EM structure of human TFIIH

Basil J. Greber[1,2], Thi Hoang Duong Nguyen[1,2,3], Jie Fang[4], Pavel V. Afonine[2], Paul D. Adams[2,5], Eva Nogales[1,2,4,6]
[1]California Institute for Quantitative Biology (QB3), University of California, Berkeley, CA, USA; [2]Molecular Biophysics and Integrative Bio-Imaging Division, Lawrence Berkeley National Laboratory, Berkeley, CA, USA; [3]Miller Institute for Basic Research in Science, University of California, Berkeley, CA, USA; [4]Howard Hughes Medical Institute, University of California, Berkeley, CA, USA; [5]Department of Bioengineering, University of California, Berkeley, CA, USA; [6]Department of Molecular and Cell Biology, University of California, Berkeley, CA, USA.

Human transcription factor IIH (TFIIH) is a 10-subunit protein complex that is part of the general transcription machinery required for the initiation of eukaryotic gene transcription by RNA polymerase II (Pol II) and additionally plays a key role in nucleotide excision DNA repair. The TFIIH core complex harbors the DNA-dependent ATPase XPB and the DNA helicase XPD, while the CDK activating kinase (CAK) subcomplex harbors CDK7, the kinase activity of which is required for regulation of transcription initiation by phosphorylation of various targets, including Pol II. Mutations in protein subunits of TFIIH have been linked to cancer and severe genetic diseases, including xeroderma pigmentosum, Cockayne syndrome, and trichothiodystrophy. In spite of its fundamental and medical importance, the high-resolution structure of intact TFIIH needed for a mechanistic understanding of its function and regulation has remained elusive. Here [1], we present the cryo-electron microscopic structure of human TFIIH at 4.4 Angstroms resolution, obtained by imaging TFIIH that has been natively purified from HeLa cells. The structure reveals the molecular architecture of the human TFIIH core complex, the detailed structures of its constituent XPB and XPD ATPases, and how the core and kinase subcomplexes of TFIIH are connected. Additionally, our structure provides insight into the conformational dynamics of TFIIH and the regulation of its activity.

Efforts to improve the resolution of the cryo-EM reconstruction to better than 4 Angstroms in order to enable building and refinement of a complete structure of TFIIH are currently underway.

[1] Greber BJ et al. (2017). The cryo-electron microscopy structure of human transcription factor IIH. Nature 549 (7672): 414-417.

31

POSTER NUMBER: 1030

Structures of nanoparticle immunogens for vaccine development

Audray Harris [1], J.R. Gallagher[1], D.M. McCraw[1], U. Torian[1], M.T. Conlon[1], M.L. Myers[1]
[1]Laboratory of Infectious Diseases, NIAID, National Institutes of Health, Bethesda, MD 20892, USA

Viral diseases affect millions of people worldwide on an annual basis. These viruses, such as influenza, are associated with human mortality and morbidity. Influenza displays antigenic variation of the major surface glycoprotein, hemagglutinin (HA), and as a consequence vaccines must be formulated each year to match both current and predicted future circulating strains. This allows the lead-time needed for vaccine production. Interestingly, there are conserved epitopes within the stem region of HA. Stem epitopes are targets for broadly neutralizing antibodies. However, one major challenge is to design influenza vaccine immunogens that can elicit antibodies to conserved epitopes. Currently, there is an increasing interest in the design and use of nanoparticles as vaccine immunogens for various diseases. However, the structure and epitope display of influenza epitopes on vaccine nanoparticles has not been explored in great detail because many structures do not form crystals that can be studied by protein x-ray crystallography. We have used cryo-electron microscopy (cryo-EM) to study the 3D structures and conserved epitope displays of different types of influenza vaccine nanoparticles. These include HA-based subunit vaccines, designed octahedral nanoparticles, and virus-like particles. Our results indicated that conserved HA stem epitope display can be modulated by both particle size and organization. In addition, different types of nanoparticles can present the HA in a prefusion state to which stem antibodies bind. However, the level of epitope accessibility could vary depending on nanoparticle design. Taken together our studies suggest that further analyses of structures of nanoparticles will aid in the design of immunogens for more efficacious influenza vaccines. The general structural design principles identified from cryo-EM studies of influenza vaccine nanoparticles could be applied to other viral systems such as RSV, Ebola, SARS coronavirus, Zika, and HIV.

McCraw, D.M., Gallagher, J.R., Harris, A.K. (2016) Clin. Vaccine Immunol. 23:483-95.

Gallagher, J.R., Harris A.K. (2016). Microsc. Micronal. (Suppl 3) 22:1112-1113.
Yassine, H.M., et al. Graham, B.S. (2015) Nat. Med. 21:1065-70.

32

POSTER NUMBER: 1036

Embracing cryoEM in drug discovery

Jun He[1], Alistair Henry[1], Paul Creek[1], Jiye Shi[2], Alastair Lawson[1], Tom Ceska[1]
[1]Structural Biology, UCB Celltech, Slough, UK; [2]CADD, UCB Pharma, Braine-l'Alleud, Belgium

CryoEM has been quickly adopted in pharmaceutical industry as an important tool to tackle proteins that are not crystallisable. Despite the successful applications to large macromolecular complex and membrane protein complex, it is highly desirable to apply this technique to lower molecular weight protein targets, which are often typical in pharmaceutical industry. Fragment antigen binding (Fab) has been described as a potential scaffold to not only enlarge the target of interest but also provide adequate structural features for successful particle alignment. Here we describe a general strategy to generate and select rigid fab-antigen complexes that are suitable for cryoEM study. We used this approach to successfully obtain atomic resolution structures of a few drug targets, ranging from 35kDa to 70kDa, which demonstrated its potential application in pharmaceutical industry.

33

POSTER NUMBER: 1031

***In-situ* structural studies of RPTPmu by cryo-electron tomography of FIB milled cells**

Christina Heroven[1,2], Christoph Hagen [3], Michael Grange [2], Kay Grünewald [2,3], Radu Aricescu[1,2]
[1]MRC Laboratory of Molecular Biology, Cambridge, UK;
[2]Division of Structural Biology, Oxford University, Oxford, UK;
[3]Heinrich Pette Institute, Leibniz Institute for Experimental Virology, Hamburg, Germany

Receptor protein tyrosine phosphatases (RPTPs) are molecules that integrate both adhesive and signaling functions and are crucial for cell-cell contact regulation. It is suggested that RPTPs have the potential to form ordered arrays of receptor molecules at cell contact sites, which has important functional implications. By using cryo-electron tomography on cells thinned by a Focused Ion Beam (FIB), we are studying the *in-situ* structure of human RPTPmu in membrane-to-membrane contacts. Here, we show how RPTPmu forms a regular array of trans-dimers at a cell contact and describe the occurrence of extensive, regularly spaced ER membrane stacks as a result of RPTPmu overexpression. With the help of sub-tomogram averaging, this data holds the potential to reveal insight into the macromolecular organization of RPTPmu in its cellular context, at high resolution.

34

POSTER NUMBER: 2033

Structural basis for protein translocation by the Vps4 AAA ATPase

Christopher P. Hill
University of Utah, Department of Biochemistry

The hexameric AAA ATPase Vps4 drives membrane fission by remodeling and disassembling ESCRT-III filaments. We have determined a 3.2 Å cryo-EM structure of Vps4 bound to an ESCRT-III peptide substrate that reveals that the peptide binds in a conformation that approximates a beta-strand, whose helical symmetry matches that of the five Vps4 subunits it contacts directly. Adjacent Vps4 subunits make equivalent interactions with successive substrate dipeptides through two distinct classes of side chain binding pockets that are formed primarily by residues of Vps4 pore loop 1. These pockets are seen to bind a wide range of residues, while main chain hydrogen bonds may help dictate substrate-binding orientation. The structure supports a "conveyor belt" model of translocation in which ATP binding allows a Vps4 subunit to join the growing end of the helix of Vps4 subunits and thereby engage the substrate. In a concerted manner, ATP hydrolysis and phosphate release promote disassembly of a Vps4 subunit from the lagging end of the Vps4 helix, thereby driving disengagement from the ESCRT-III substrate.

35

POSTER NUMBER: 1032

Cryo-EM tomography and subtomogram averaging of COPII assembled on membranes

Joshua Hutchings[1], Giulia Zanetti[1]
[1]Institute of Structural and Molecular Biology, Birkbeck, Malet St, London, WC1E 7HX

Coat protein complex II (COPII) is an archetypal Eukaryotic coat protein. It governs vesicular transport from the Endoplasmic Reticulum (ER) to the Golgi apparatus, a key step in the secretory pathway. It assembles in a hierarchical manner to selectively incorporate proteins into vesicles, and it organises into an "inner" and "outer" coat layer. COPII is deployed to accommodate a diverse range of cargo proteins, including smaller soluble proteins and larger structures like procollagen. A number of genetic diseases arising from mutants of COPII and its effectors have been identified, pertaining to compromised trafficking of procollagen and other large cargo proteins. Despite its fundamental role, the structural and regulatory mechanisms governing COPII adaptability remain enigmatic.
We chose to address the structural basis for the versatile deployment of COPII in ER-to-Golgi membrane trafficking using cryo-electron tomography and sub-tomogram averaging.
We have reconstituted COPII vesicle formation using recombinantly purified proteins and giant unilamellar vesicles (GUVs). The reconstitution yielded elongated tubular membranes wrapped in a helical lattice, corresponding to the COPII inner coat as seen previously (1). We acquired a large cryo-tomography dataset of COPII-coated tubules. We used the dose-symmetric scheme on a Titan Krios equipped with a K2 detector and post-column energy filter, and a pixel size of 1.3Å. Dose-compensation and 3D-CTF correction have been applied to preserve the high-resolution features of the tomograms.
Using these developments, our current subtomogram averaging has yielded a 6.8Å reconstruction of the COPII inner coat assembled on membranes, allowing us to obtain new insights into the supramolecular assembly of COPII in action. For example, the hitherto unseen membrane insertion by a single helix that underpins COPII assembly is already partially resolved in our reconstructions. We are in the process of optimising our alignments to obtain a near-atomic resolution structure of the coat. Moreover, inspired by knowledge of existing genetic diseases, candidates for mutagenesis are being identified to elaborate a model for the molecular mechanisms of COPII deployment in ER-to-Golgi transport.

References:
1 - doi:10.7554/eLife.00951.

36

POSTER NUMBER: 2001

CryoEM Structure of the Human Sodium-Coupled Acid-Base Transporter NBCe1

Kevin W. Huynh[1,2*], Jiansen Jiang,[2,3*], Natalia Abuladze[1*], Kirill Tsirulnikov[1*], Liyo Kao[1], Xuesi Shao[4], Debra Newman[1], Rustam Azimov[1], Alexander Pushkin[1], Z. Hong Zhou[2,3], and Ira Kurtz[1,5]
[1]Department of Medicine, Division of Nephrology, David Geffen School of Medicine, University of California, Los Angeles, California, USA; [2]California NanoSystems Institute, University of California, Los Angeles, California, USA; [3]Department of Microbiology, Immunology & Molecular Genetics, University of California, Los Angeles, California, USA; [4]Department of Neurobiology, University of California, Los Angeles, California, USA; [5]Brain Research Institute, University of California, Los Angeles, California, USA
*These authors equally contributed to this work

Na^+-coupled acid-base membrane transport proteins play an essential role in human biology. Dysfunction of these symporters and exchangers has been linked to cancer, heart and brain disease. Despite their importance, high-resolution structures of mammalian Na^+-coupled acid-base transporters are not available. NBCe1, a Na^+-coupled CO_3^{2} symporter, functions in multiple organs and its mutations cause blindness, growth abnormalities, migraines, impaired cognitive function and abnormal blood chemistry. Here, we have determined the structure of human NBCe1 at 3.9 Å resolution by cryo electron microscopy. In solution, NBCe1 exists as a homodimer whose structure resembles the iconic "double-headed eagle", with a gate domain, core domain, large extracellular loop and cytoplasmic region of each NBCe1 monomer corresponding to the eagle's body, wing, head and foot, respectively. Our atomic model and functional mutagenesis revealed the ion accessibility pathway and the ion coordination site, the latter containing residues involved in human disease-causing mutations. Using this structural and functional information, we identified a small number of residues within the ion coordination site whose modification transformed NBCe1 into an anion exchanger. Our data suggest that symporters and exchangers utilize comparable transport machinery and that subtle differences in their substrate binding regions have very significant effects on their transport mode.

37

POSTER NUMBER: 2002

Electron microscopy and integrative modeling shed light on the mechanisms of transcription initiation

Chunli Yan[2], Yan Han[1], Yuan He[1], K. Nguyen[3], Eva Nogales[3], Ivaylo N. Ivanov[2*]
[1]Department of Molecular Biosciences, Northwestern University, Evanston, IL
[2]Department of Chemistry, Georgia State University, Atlanta, GA
[3]Department of Molecular and Cell Biology, University of California, Berkeley, CA

Advances in cryo-EM have opened up new avenues to structurally define biomolecular assemblies. To arrive at psedoatomic models, it is necessary to employ integrative computational modeling. Here, we illustrate the success of this hybrid approach in describing pre-initiation complexes of eukaryotic RNA Polymerases (Pol I and Pol II). During initiation Pol II and general transcription factors (GTFs) assemble into a pre-initiation complex (PIC). The initial closed promoter complex (CC) transitions into an open complex (OC), in which the melted template DNA is inserted into the Pol II active site. The OC is eventually converted into an initial transcribing complex (ITC). Atomic resolution description of this complex machinery had proved elusive. Combining cryo-EM with state-of-the-art computation allowed visualization of the Pol II PIC at all major stages of transcription initiation. Comparisons of the functional states provided new insights into the processes of DNA engagement, promoter melting and transcription bubble stabilization. By contrast, Pol I initiation depends on the interaction of Pol I with the GTF Core Factor (CF). Visualizing the Pol I PIC showed that Pol I and promoter DNA are a preconditioned in an elongation competent form. The structures revealed a bipartite architecture of the CF. Here the Pol I PIC was captured in three distinct functional states. Comparison to the Pol II system suggests that a ratchet motion of the CF bound to the upstream DNA facilitates promoter melting in an ATP-independent manner.

1. He et al. Structural basis of transcription promoter opening using single particle cryo-EM. *Nature* (2016) 533, 359-365
2. Han et al. Structural mechanism of ATP-independent transcription initiation by RNA polymerase I. *eLife* (2017) 6, e27414

38

POSTER NUMBER: 2003

Structure of the human MHC-I peptide-loading complex

Andreas Blees[1+], Dovile Januliene[2], Tommy Hofmann[3], Nicole Koller[1], Carla Schmidt[3], Simon Trowitzsch[1*], Arne Moeller[2*], Robert Tampé[1*]
[1] Goethe University Frankfurt, Germany; [2] Max Planck Institute of Biophysics, Germany; [3] Martin Luther University Halle-Wittenberg, Germany
+These authors contributed equally to this work
*Corresponding Authors

The peptide-loading complex (PLC) is a transient, endoplasmic reticulum (ER) resident, multisubunit membrane protein complex, essential for establishing a hierarchical immune response. The PLC coordinates peptide translocation into the ER with loading and editing of major histocompatibility complex class I molecules (MHC-I). After final proofreading in the PLC, stable peptide/MHC-I complexes are released to the cell surface to evoke a T-cell response against infected or malignant cells. Sampling of different MHC-I allomorphs requires the precise coordination of seven different subunits in a single macromolecular assembly, including the transporter associated with antigen processing TAP1/2, the oxidoreductase ERp57, the MHC-I heterodimer and the chaperones tapasin and calreticulin.
In order to decipher the molecular architecture of this intrinsically dynamic multisubunit complex, we isolated human PLC from Burkitt's lymphoma cells using an engineered viral inhibitor as a bait and determined the structure of native PLC by electron cryo-microscopy (cryo-EM). Two ER-lumenal editing modules composed of tapasin, calreticulin, ERp57 and MHC-I are centered around TAP in a pseudo-symmetric orientation. A multivalent chaperone network within and across the editing modules establishes the proofreading function at two lateral binding platforms for MHC-I clients. The lectin-like domain of calreticulin senses the MHC-I glycan, while the P-domain reaches over the MHC-I peptide-binding pocket towards ERp57. This arrangement allows tapasin to exert peptide editing by clamping MHC-I. The translocation pathway of TAP opens out into a large ER-lumenal cavity, confined by the membrane entry points of tapasin and MHC-I. Two lateral windows channel the antigenic peptides to MHC-I. PLC captured at distinct assembly states provide mechanistic insight into MHC-I recruitment and release. Our work

defines the molecular symbiosis of an ABC transporter and an ER-chaperone network in MHC-I assembly and illuminates the onset of an adaptive immune response.

39

POSTER NUMBER: 2004

Structural Basis of Mitochondrial Receptor Binding and GTP-Driven Conformational Constriction by Dynamin-Related Protein 1

Raghav Kalia[1,2,3,5], Ray Yu-Ruei Wang[1,3,4], Ali Yusuf[1,3], Paul V. Thomas[1,3,5], David A. Agard[1,3,4], Janet M. Shaw[2,4] and Adam Frost[1,2,3,5]
[1]Department of Biochemistry and Biophysics, University of California, San Francisco, San Francisco, CA 94158 USA
[2]Department of Biochemistry, University of Utah, Salt Lake City, UT, 84112 USA
[3]California Institute for Quantitative Biomedical Research, San Francisco, CA 94158 USA
[4]Howard Hughes Medical Institute
[5]Chan Zuckerberg Biohub, San Francisco, CA 94158 USA

Mitochondrial inheritance, genome maintenance, and metabolic adaptation all depend on organelle fission by Dynamin-Related Protein 1 (Drp1) and its mitochondrial receptors (1). Drp1 receptors include the paralogs Mitochondrial Dynamics 49 and 51 (MiD49/MiD51) and Mitochondrial Fission Factor (MFF), but the mechanisms by which these proteins recruit Drp1 and regulate its activities are unknown (2,3). Here we present a 4Å resolution cryo-EM structure of human, full-length Drp1 bound to MiD49 and an analysis of structure- and disease-based mutations. We report that GTP binding allosterically induces a remarkable elongation and rotation of the G-domain, Bundle-Signaling Element (BSE) and connecting hinge loops of Drp1. In this nucleotide-bound conformation, a distributed network of multivalent interactions promotes Drp1 copolymerization into a linear filament with MiD49, MiD51 or both. Subsequent GTP hydrolysis and exchange within the filament leads to receptor dissociation, shortening through disassembly, and concomitant curling of Drp1 oligomers into closed rings. The dimensions of the closed Drp1 rings are consistent with Drp1-constricted mitochondrial tubules observed in human cells (4). These structures are the first views of full-length, receptor- and nucleotide-bound dynamin-family GTPases and—in comparison with nucleotide-free crystal structures—teach us how these molecular machines perform mechanical work through nucleotide-driven allostery.

References

(1) Mishra, P. and Chan, D.C. *Nat. Rev. Mol. Cell Biol.* **15**, 634-646 (2014).
(2) Palmer, C.S. *et al. EMBO Rep.* **12**, 565-573 (2011).
(3) Gandre-Babbe, S. and van der Bliek, A.M. *Mol. Biol. Cell* **19**, 2402-2412 (2008).
(4) Lee, J.E. *et al. Nature* **540**, 139-143 (2016).

Funding
The Howard Hughes Medical Institute, National Institutes of Health and the Searle Scholars Program.

40

POSTER NUMBER: 2005

Molecular mechanism of RNA hairpin-stabilized pause and pause inhibition by elongation factors NusG and RfaH during bacterial transcription

Jin Young Kang[1], Tatiana Mishanina[2], Rachel A. Mooney[2], Yuri Nedialkov[3], Irina Artsimovitch[3], Robert Landick[2], and Seth A. Darst[1]
[1]The Rockefeller University, 1230 York Avenue, New York, NY, USA
[2]Department of Biochemistry, University of Wisconsin-Madison, Madison, WI, USA
[3]Department of Microbiology, The Ohio State University, 484 West 12th Avenue, Columbus, OH, USA

RNA polymerase (RNAP) pauses at every ~100 base pairs during transcription in the model bacterium, *E. coli*. In bacteria, nascent RNA structures thought to form within the RNAP RNA exit channel can increase pause lifetimes ten-fold or more. Despite its important regulatory roles in transcription initiation, elongation, RNA folding and termination, the underlying molecular mechanism of transcriptional pausing has remained elusive. To probe the pause mechanism, we determined a 3.8 Å CryoEM structure of an *E. coli his* pause elongation complex (*his*PEC), which is stabilized by a nascent RNA hairpin in the leader sequence of the his operon. We found that i) the double-stranded A-form RNA hairpin stem forms within the RNAP RNA exit channel, ii) the DNA:RNA hybrid is trapped in a distinctive half-translocated conformation, and iii) mobile domains of the RNAP (dock, clamp, jaw, shelf and SI3) rearrange in a concerted manner to maintain the hairpin-stabilized paused state. These structural features provide molecular insight into transcriptional pausing and its stabilization by the RNA hairpin. To understand how pausing is regulated, we determined CryoEM structures of elongation complexes bound by two antipausing factors, NusG and RfaH, at 3.7 Å and 3.5 Å resolution, respectively. Although NusG is a ubiquitous pausing inhibitor, RfaH associates with RNAP by sequence-specific recognition of the "*ops*" element. NusG and RfaH bind to the ß protrusion, ß gate loop, and ß' clamp helices of RNAP, stabilizing RNAP with a closed-clamp conformation that resembles the on-pathway elongation complex. In the RfaH-elongation complex structure, RfaH interacts sequence-specifically with *ops*, which forms a short hairpin in the non-template DNA. Based on our results, we propose that the pause RNA hairpin stabilizes global conformational changes in the RNAP that secures the paused state whereas NusG and RfaH enhance transcription elongation by preventing these conformational changes.

41

POSTER NUMBER: 2006

Cryo-EM structure of the full human PRC2 and regulation by AEBP2 and JARID2

Vignesh Kasinath[1,2], Marco Faini[3], Dvir Reif[1], Xinyu Ashlee Feng[4], Goran Stjepanovic[1,2], Simon Poepsel[1,2], Ruedi Aebersold[3,5] and Eva Nogales[1,2,6*]
[1]QB3 Institute, Department of Molecular and Cell Biology, University of California, Berkeley, California 94720, USA; [2]Molecular Biophysics and Integrated Bioimaging Division, Lawrence Berkeley National Laboratory, USA; [3]Department of Biology, Institute of Molecular Systems Biology, ETH Zurich, 8093, Zurich, Switzerlandl; [4]Department of Chemistry, University of California Berkeley, USA; [5]Faculty of Science, University of Zurich, Switzerland; [6]Howard Hughes Medical Institute, University of California, Berkeley, California 94720, USA
* Corresponding author: enogales@lbl.gov

The ability to establish and maintain cellular identity is critical for development in multicellular organisms. Central to this process is PRC2 (Polycomb repressive complex 2), a key epigenetic regulator responsible for gene silencing by catalyzing histone H3K27 methylation leading to heterochromatin formation. Here we report cryo-EM structures of the full human PRC2 in several functional states, in complex with its cofactors JARID2 and AEBP2. Both cofactors mimic the binding of histone H3 tails. JARID2 is methylated at lysine 116 and stimulates PRC2 through its interaction with EED, while AEBP2 interacts with the RBAP48 site mimicking unmodified histone H3. SUZ12 extends through the complex, establishing interactions with all other PRC2 subunits. A SUZ12-AEBP2-JARID2 sub-complex links the EZH2-EED lobe to the SUZ12-RBAP48 lobe stabilizing the full complex. Our structural analysis reveals the differences in the stimulatory effect of JARID2 and AEBP2 binding, while highlighting the importance of their synergistic effect on PRC2 activity.

42

POSTER NUMBER: 2007

Architecture of Volume-Regulated Anion Channels

Jennifer M Kefauver[1,2], Gunhee Hong[2], Christopher S Crowley[1], Adrienne E Dubin[2], Christopher A Cottrell[1], Ardem Patapoutian[2*], Andrew B Ward[1*]
[1]The Scripps Research Institute, Department of Integrative Structural and Computational Biology; [2]The Scripps Research Institute, Department of Neuroscience
*Corresponding authors

Every vertebrate cell must be able to respond rapidly to changes in its volume to protect itself from bursting. The Volume Regulated Anion Channel (VRAC) plays an important and well-characterized role in regulatory volume decrease, releasing chloride ions and small osmolytes to restore osmotic balance in response to hypo-osmotic stress. Although currents produced by VRAC were first observed almost 30 years ago and these currents are observed in nearly every vertebrate cell, it was only recently shown that the molecular components of VRAC comprised of five Leucine-Rich Repeat-Containing 8 (LRRC8) proteins (A-E). LRRC8 proteins have four predicted N-terminal transmembrane domains and 17 leucine-rich repeats at their cytosolic C-termini. Whole-cell VRAC currents are produced by obligate heteromers composed of the LRRC8A subunit and one or more other LRRC8 family members. Our data reveal that VRACs are hexamers; however, this implies a possible compositional landscape of over one thousand combinations and nothing is known about which ones occur *in vivo*. To eliminate the problem of compositional heterogeneity, we have used CRISPR to produce LRRC8B-E$^{-/-}$ HEK 293F suspension cells and we overexpress affinity-tagged LRRC8A to produce LRRC8A homomeric channels, which have been shown to be functional channels when reconstituted in lipid bilayers. We present a symmetric subnanometer reconstruction of a homomeric LRRC8A channel that gives us insights into the architecture of VRACs.

This work was supported by NIH F31 NS093778-3 awarded to J.M.K., NIH R01 1NS083174 awarded to A.P., and a Ray

Thomas Edwards Foundation Grant awarded to A.B.W.

43

POSTER NUMBER: 2008

Understanding morphological changes accompanying sporulation in *B. subtilis* using in situ cryo-electron tomography

Kanika Khanna[1], Javier Lopez-Garrido[1], Ziyi Zhao[1], Yuan Yuan[1], Kit Pogliano[1], Elizabeth Villa[1*],
[1]Division of Biological Sciences, University of California, San Diego
*Corresponding Author

Engulfment is a hallmark of the developmental pathway that culminates in the production of dormant endospores by Gram-positive bacteria providing an ideal system for the study of how bacterial cells localize and move macromolecules and how they spatially and temporally regulate peptidoglycan degradation and synthesis. Recent technological and image processing advances place cryo-electron microscopy methods in a unique position to image cells in a frozen-hydrated state that reveals the structure and organization of macromolecules in their cellular context. The modality of cryo-electron tomography (CET) can be used to obtain 3D images of cells at molecular resolution and focused-ion beam (FIB) milling at cryogenic temperatures provides a suitable way to micromachine samples to a thickness suitable for CET. Here, we explore the capability of cryo-FIB milling coupled with CET to study the developmental process of sporulation in *Bacillus subtilis*. We aim to achieve a mechanistic model of engulfment, focusing on how peptidoglycan remodeling and chromosome translocation contribute to the topology of engulfment. We propose a model for engulfment that entails the coordination of PG synthesis and degradation between the two cells of the sporangium, and in which the chromosome inflates the forespore like air in a balloon. Our images also show various as yet unidentified filamentous and globular structures that we are characterizing using computational approaches such as subvolume averaging and template matching. Thus by combining FIB-CET with in silico image analysis we aim to create a visual atlas documenting the cellular transformations of sporulation at molecular resolution.

44

POSTER NUMBER: 2009

Mechanism of cellular recognition and entry by porcine circovirus 2

Somali Dhindwal[1], Shanshan Feng[1,2], Bryant Avila[1], Reza Khayat[1,2*],
[1]City College of New York, Department of Chemistry and Biochemistry; [2]PhD programs in Biochemistry and Chemistry, The Graduate Center, City University of New York, NY USA
*Corresponding Author

Swine pathogens are known to infect and inflict their pathogenic phenotype in humans. Porcine circovirus 2 (**PCV2**) is a non-enveloped immunosuppressive virus responsible for eradicating swine farms throughout the planet. PCV2 is promiscuous in that it infects and destroys nearly every tissue and organ within its natural host, infects and induces immunosuppression in multiple species of the animal kingdom, and is capable of infecting and replicating in human cells in culture -properties that give PCV2 the propensity of becoming zoonotic and a deterrent to xenotransplantation. Xenotransplantation from swine to human is becoming increasingly attractive due to the shortage of organs from human donors. Understanding the detailed mechanisms by which PCV2 interacts with its cellular hosts is of significant importance for the development of therapeutics to intervene with PCV2 infection. Moreover, PCV2 is the smallest known pathogenic virus and thus provides the opportunity to understand the minimal components needed by an autonomously replicating virus. The presentation discusses our findings in the cellular recognition and entry by PCV2.

45

POSTER NUMBER: 2010

***CryoFIT*: Flexible Fitting of Biomolecule Structures from Cryo-EM Maps with User Friendly Molecular Dynamics Simulation**

Doo Nam Kim[1], Serdal Kirmizialtin[2], Karissa Sanbonmatsu[1*],
[1]Los Alamos National Laboratory; [2]New York University at Abu Dhabi
*Corresponding Author

With many crystallography groups shifting to cryogenic electron microscopy (cryo-EM), there is a high demand for software that can produce atomistic models for cryo-EM datasets. However, only molecular dynamics (MD) simulation experts have been able to fit challenging targets due to complicated MD steps such as specifically preparing pdb file for a certain force field, generating topology, minimization, neutralization, generating constraints, running MD simulation and analysis. Therefore, we incorporated our automated cryo-EM density fitting method (*cryoFIT*) into a widely used software suite for macromolecular structure determination (*PHENIX*) and now it is publicly available to be used by any scientists with valid *PHENIX* academic or industry license. *CryoFIT* produces all-atom models highly consistent with the EM density and is being used to construct models for many functional complexes. Three key advantages of *cryoFIT* are its speed, preservation of stereochemistry and user-friendliness. Its fast speed allows cryo-EM scientists to run on a single laptop (several hours for a ribosome modeling without the need for GPU) instead of thousands of cores. To preserve stereochemistry, secondary and tertiary contact potentials are automatically specified to place additional restraints on contacts within a cutoff keeping them intact during the fitting process. Convenient automatic installation, *PHENIX*-GUI based running, automatic option assignments and continuous running of all steps make this program very user-friendly. To show concrete examples, application of this program into the human ribosome and nucleosome will be presented.

Funding
NIH

46

POSTER NUMBER: 2034

Structural and Functional Anatomy of a Nuclear Pore Complex

Seung Joong Kim[1*], Javier Fernandez-Martinez[2*], Ilona Nudelman[2*], Yi Shi[3,13*], Wenzhu Zhang[3*], Barak Raveh[1], Thurston Herricks[4], Brian D. Slaughter[5], Joanna Hogan[6], Paula Upla[7], Ilan E. Chemmama[1], Riccardo Pellarin[1,8], Ignacia Echeverria[1], Manjunatha Shivaraju[5], Azraa S. Chaudhury[2], Junjie Wang[3], Rosemary Williams[2], Jay R. Unruh[5], Charles H. Greenberg[1], Erica Y. Jacobs[3], Zhiheng Yu[9], M. Jason de la Cruz[9], Roxana Mironska[2], David L. Stokes[7], John D. Aitchison[4,10], Martin F. Jarrold[6], Jennifer L. Gerton[5], Steven J. Ludtke[11†], Christopher W. Akey[12†], Brian T. Chait[3†], Andrej Sali[1†], and Michael P. Rout[2†]
[1]Dept. of Bioengineering and Therapeutic Sciences, Dept. of Pharmaceutical Chemistry, and California Institute for Quantitative Biosciences, University of California, San Francisco, CA 94158; [2]Laboratory of Cellular and Structural Biology, The Rockefeller University, New York, NY 10065; [3]Laboratory of Mass Spectrometry and Gaseous Ion Chemistry, The Rockefeller University, NY 10065; [4]Institute for Systems Biology, 401 Terry Ave. N., Seattle, WA 98109; [5]Stowers Institute for Medical Research, Kansas City, MO 64110; [6]Department of Chemistry, Indiana University, Bloomington, IN 47405; [7]Skirball Institute and Department of Cell Biology, New York University School of Medicine, NY 10016; [8]Current address: Structural Bioinformatics Unit, Institut Pasteur, CNRS UMR 3528, Paris, France; [9]Janelia Research Campus, Howard Hughes Medical Institute, 19700 Helix Drive, Ashburn, Virginia, 20148; [10]Center for Infectious Disease Research, Seattle, WA; [11]National Center for Macromolecular Imaging, Verna and Marrs McLean Department of Biochemistry and Molecular Biology, Baylor College of Medicine, 1 Baylor Plaza, Houston, TX 77030; [12]Dept. of Physiology and Biophysics, Boston University School of Medicine, 700 Albany Street, Boston, MA 02118; [13]Current address: Department of Cell Biology, University of Pittsburgh School of Medicine, Pittsburgh, PA, 15260.
*These authors contributed equally to this work and their order was determined through a random selection process
†Co-corresponding authors

Despite the central role of Nuclear Pore Complexes (NPCs) as gatekeepers of RNA and protein transport between the cytoplasm and nucleoplasm, their large size and dynamic nature have impeded a full structural and functional elucidation. Here, we have determined a subnanometer precision structure for the entire 552-protein yeast NPC by satisfying diverse data, including primarily stoichiometry, a cryo-electron tomography map, and chemical cross-links. The structure reveals the NPC's functional elements in unprecedented detail. The NPC is built of sturdy diagonal columns to which are attached connector cables, imbuing both strength and flexibility, and tying together all other elements of the NPC, including membrane-interacting regions and RNA processing platforms. Inwardly-directed anchors create a high density of transport factor-docking Phe-Gly repeats in the NPC's central channel, organized in distinct functional units. Taken together, this integrative structure allows us to rationalize the architecture, transport mechanism, and evolutionary origins of the NPC.

47

POSTER NUMBER: 2011

The Ins and Outs of Benchmarking your cryo-TEM

Laura Y. Kim, Edward T. Eng, William J. Rice, Anchi Cheng, Ashleigh M. Raczkowski, Kelsey D. Jordan, Daija Bobe, Venkat Dandey, Zhening Zhang, Alex Hui Wei, Bridget Carragher, Clinton S. Potter.
Simons Electron Microscopy Center, New York Structural Biology Center, New York, NY, USA

With the increasing popularity and demand for cryo-EM as a method to acquire high-resolution 3D information from biological macromolecules, there is a need for researchers to be able to readily and easily benchmark the resolution limits of their instrumentation in order to assess the quality of the data acquired from said microscope. One of the key components of benchmarking a TEM is access to a biological specimen that can be easily prepared for cryo-EM data collection and processing. This is especially important for cryo-EM facilities, where there are several microscopes in use. Unfortunately, benchmarking endeavors can become quite laborious as there is no "industry standard" for biological specimen, and also because there is so much intrinsic variability at the specimen purification, grid preparation, data collection and image processing levels. Here we present the single particle workflow of a 2.5 Å aldolase structure obtained at the Simons Electron Microscopy Center to highlight our efforts to establish a standardized pipeline. We also include details on what works, what does not work, and tips to improve the overall workflow. The goal of this workflow is to help microscopists in the cryo-EM field benchmark the quality of their TEM.

48

POSTER NUMBER: 2012

Structural insights into the SARS Spike-ACE2 receptor complex

Robert N. Kirchdoerfer[1], Nianshuang Wang[2], Jesper Pallesen[1], Daniel Wrapp[2], Hannah L. Turner[1], Christopher A. Cottrell[1], Kizzmekia S. Corbett[3], Barney S. Graham[3], Jason S. McLellan[2], Andrew B. Ward[1]
[1]Department of Integrative Structural and Computational Biology, The Scripps Research Institute
[2]Department of Biochemistry and Cellular Biology, Geisel School of Medicine at Dartmouth
[3]Viral Pathogenesis Laboratory, National Institute of Allergy and Infectious Disease

SARS coronavirus emerged in 2002 as a highly transmissible human pathogen causing numerous hospitalizations and associated deaths. Key to viral entry into host cells and host species tropism is the viral spike protein, which binds to host protein receptors and mediates fusion of the viral and host membranes. Previous structural characterizations of coronavirus trimeric ectodomains have been limited to the spike alone or in complex with neutralizing antibodies. Here, we present a cryoEM analysis of the SARS CoV spike-human ACE2 receptor interactions. These structures reveal that the spike binds to ACE2 using an outwards S1 C-terminal domain conformation. Even in the presence of an excess of ACE2, a substoichiometric amount of ACE2 is observed to bind the spike even though other C-terminal domains adopt outward-facing conformations available to bind receptor. These studies suggest that rather than inducing large conformational changes, ACE2 samples available C-terminal domain conformations for binding.

This work is supported by NIH/NIAID K99AI123498 and R01AI127521.

49

POSTER NUMBER: 2013

***Staphylococcus aureus* pathogenicity islands use competing scaffolding proteins for the molecular pirating of bacteriophage capsids**

James L. Kizziah[1], Altaira D. Dearborn[2], Erin A. Wall[3], Laura Klenow[3], Laura K. Parker[3], Keith A. Manning[1], Michael S. Spilman[4], John M. Spear[5], Gail E. Christie[3], Terje Dokland[1]
[1]Department of Microbiology, University of Alabama at Birmingham, Birmingham, AL 35294, USA; [2]Protein Expression Laboratory, NIAMS, NIH, Bethesda, MD 20892, USA; [3]Department of Microbiology and Immunology, Virginia Commonwealth University, Richmond, VA 23298, USA; [4]Direct Electron, LP, San Diego, CA 92128, USA; [5]Biological Science Imaging Resource, Florida State University, Tallahassee, FL 32306, USA

Staphylococcus aureus is a Gram-positive bacterium and an opportunistic human pathogen that causes dermal and systemic infections. Transfer of genes, including genes for virulence factors, among *S. aureus* usually occurs via transduction by bacteriophages, some of which can also de-repress *S. aureus* pathogenicity islands (SaPIs). SaPIs are mobile genetic elements that can encode several toxins, but are normally stably integrated into the genomes of *S. aureus*. Upon infection of the host cell by specific helper phages, SaPIs excise, replicate, and package themselves into phage-like particles commandeered from the helper phage through the process of "molecular piracy". Upon cell lysis, SaPI particles are released with the phage and infect new hosts, disseminating themselves and any virulence factors they encode. In this study, we investigated one mechanism by which SaPIs redirect the assembly of helper phage capsids to make smaller capsids capable only of accommodating the smaller SaPI genome. Using cryo-electron microscopy, we determined near-atomic resolution structures of the procapsids of SaPI1 and one of its helper phages, called 80α. These structures suggest that the 80α scaffolding protein, which is required for capsid assembly, and the SaPI1 protein CpmB, which is required for capsid size redirection, compete for the same binding site on the capsid protein, and that the smaller CpmB protein constrains the angles among capsomers to force the assembling capsid to adopt the smaller size usable only by SaPI1. Additionally, the structures show for the first time the detailed interactions between the scaffolding and capsid proteins in a bacteriophage, and along with our genetic probing of these interactions, shed light on the process of capsid assembly.

50

POSTER NUMBER: 1038

Polymerization of CTP synthase homologs in different filament geometries generates regulatory diversity

E.M. Lynch[1], D.R. Hicks[1,2], M. Shepherd[3], J.A. Endrizzi[4], A. Maker[1], J.M. Hansen[1,5], R.M. Barry[6], Z. Gitai[6], E.P. Baldwin[4], J.M. Kollman[1]
[1]Department of Biochemistry, University of Washington, Seattle , WA, [2]Graduate Program in Molecular and Cellular Biology, University of Washington, Seattle, WA, [3]Department of Biology, McGill University, Montreal, QC, [4]Department of Molecular and Cellular Biology, University of California, Davis, Davis, CA, [5]Graduate Program in Biological Physics, Structure, and Design, University of Washington, Seattle, WA, [6]Department of Molecular Biology, Princeton University, Princeton, NJ

CTP synthase (CTPS) is the best-characterized member of a growing class of novel metabolic filaments that organize and regulate cellular enzyme activity. CTPS is a universally conserved and essential metabolic enzyme, responsible for catalyzing the rate-limiting step in CTP biosynthesis - conversion of UTP to CTP in an ATP-dependent process. CTP is critical to cellular function, being a fundamental component of DNA and RNA, and also key in the synthesis of membrane phospholipids. Further, all four ribonucleotides influence CTPS activity, making it an important regulatory hub of nucleotide metabolism. CTPS in cells polymerizes into micron-scale filaments from bacteria to humans. We recently showed that the bacterial and human CTPS filaments are structurally distinct - in both cases the polymers take advantage of conserved allosteric regulation of the enzyme, but the human filament stabilizes an active conformation and the bacterial enzyme stabilizes an inactive conformation. Here, we show that polymerization of human CTPS1, an isoform upregulated during cellular proliferation, is regulated by post-translational modification, and present the high resolution cryo-EM structure of the filament in the assembled active conformation. We also find that CTPS2, the housekeeping isoform of the enzyme, assembles filaments that can interchange between active and inactive conformations in the polymer, properties that are conserved with the yeast enzyme. The variation in filament geometries formed by different CTPS homologs illustrates the diversity of regulatory strategies that can be built from polymerization-based regulation of enzymes.

51

POSTER NUMBER: 2035

Structure and Regulation of Respiratory Syncytial Virus RNA Polymerase

Dongdong Cao[1#], Yunrong Gao[1#], Bo Liang[1*]
[1]Department of Biochemistry, Emory University School of Medicine, Atlanta, GA 30322
*Corresponding Author

Cryo-electron microscopy (cryo-EM) is especially well suited for molecular systems traditionally challenging for structural characterization, including membrane proteins and large and heterogeneous assemblies. One such challenging system is the RNA synthesis machinery of a class of pathogenic and sometimes deadly non-segmented negative-sense (NNS) RNA viruses, including rabies, measles, Ebola, Marburg, and respiratory syncytial virus (RSV). Central to the life of these viruses is RNA synthesis, which is carried out by the RNA polymerase (the multifunctional enzyme). However, the structural basis of the RNA synthesis machineries of these viruses remains largely unclear. The Liang laboratory is dedicated to understanding the structure and function of the RNA synthesis machinery of RSV, the top leading cause of severe pediatric respiratory tract diseases worldwide. The RNA synthesis by RSV RNA polymerase is essential for viral pathogenesis in infants and children. The Liang laboratory will establish a RNA synthesis platform for RSV, elucidate how this RNA synthesis machine functions, and identify potential antiviral therapeutic targets for more effective treatment. Our immediate research goal is to decipher the molecular architecture of the RSV RNA synthesis machine using single particle cryo-EM and x-ray crystallography. This could lead to the development of effective antiviral drugs to block the RSV activity. Such drugs would result in a substantial reduction in serious RSV infections in infants and children and provide a major achievement for human health.

52

POSTER NUMBER: 2015

Near-atomic resolution cryo-EM for a small protein on a symmetric imaging scaffold

Yuxi Liu[1,*] Shane Gonen[2,*], Tamir Gonen[2,3], Todd O. Yeates[1,4,5$]
[1]UCLA Department of Chemistry and Biochemistry [2]Howard Hughes Medical Institute, Janelia Research Campus [3]Howard Hughes Medical Institute, Departments of Physiology and Biological Chemistry, David Geffen School of Medicine, UCLA [4]UCLA-DOE Institute for Genomics and Proteomics [5]UCLA Molecular Biology Institute
*Equal author contributions
$Corresponding author: Todd O. Yeates, UCLA Dept. of Chemistry and Biochemistry, Los Angeles, CA 90095-1569

With several critical technical advancements, current cryo-electron microscopy (cryo-EM) methods can produce images of large protein assemblies and macromolecular complexes at atomic level detail. However, atomic resolution cryo-EM remains challenging or impossible for targets below about 50 kDa. The present work aims to develop a facile molecular scaffolding system to enable the assembly of small and asymmetric proteins into large and symmetric nanoparticles to overcome the cryo-EM lower size limit. Using a continuous alpha-helical linker, we connect a small 17 kDa protein (DARPin) to a previously designed symmetric protein assembly (*1*). We are able to identify and solve the structure of the DARPin in the resulting construct at near-atomic resolution by single-particle cryo-EM. Our result demonstrates that the alpha-helical linker is a sufficiently rigid attachment between the DARPin and the symmetric assembly, allowing the cryo-EM analysis of a protein otherwise too small. Additionally, the amino acid sequence of a DARPin can be selected for tight binding affinity to various other proteins (2), which can assemble onto the current construct through their interaction with the DARPin. Therefore, we propose that the current construct can be expanded into a general scheme for imaging diverse macromolecules in the future, potentially broadening the application of cryo-EM to proteins of typical size in the cell.

References

1. N. P. King *et al.*, Accurate design of co-assembling multi-component protein nanomaterials. *Nature*. **510**, 103-108 (2014).
2. H. K. Binz, M. T. Stumpp, P. Forrer, P. Amstutz, A. Plückthun, Designing Repeat Proteins: Well-expressed, Soluble and Stable Proteins from Combinatorial Libraries of Consensus Ankyrin Repeat Proteins. *J. Mol. Biol.* **332**, 489-503 (2003).

53

POSTER NUMBER: 2016

Structural Analysis Reveals the Features of Ribosome Assembly Factor Nsa1/WDR74 Important for Localization and Interaction with the AAA-ATPase Rix7/NVL2

Yu-Hua Lo, Erin M. Romes, Monica C. Pillon, Mack Sobhany, and Robin E. Stanley*
From the Signal Transduction Laboratory, National Institute of Environmental Health Sciences, National Institutes of Health, Department of Health and Human Services, 111 T. W. Alexander Drive, Research Triangle Park, North Carolina 27709, USA

Ribosome assembly is a complex process that requires hundreds of essential assembly factors, including Rix7 (NVL2 in mammals) and Nsa1 (WDR74 in mammals). Previous studies in *Saccharomyces cerevisiae* suggest that Rix7 mediates the release of Nsa1 from nucleolar pre-60S particles however the underlying mechanisms of this release are unknown. Here, we solved the first crystal structure of *S. cerevisiae* Nsa1 at 1.3 Å resolution and used small angle X-ray scattering to determine the solution structure of full length Nsa1. Through a series of co-immunoprecipitation and immunofluorescence assays with the mammalian homologues we found that the N-terminal WD40 domain of WDR74 is required for interaction with NVL2 while the flexible C-terminal tail is required for nucleolar localization. Intriguingly we found that the interaction between WDR74 and NVL2 is not dependent upon RNA or pre-60S particles demonstrating that WDR74 and NVL2 can associate with one another both on and off pre-60S particles. We further show that WDR74 associates with the D1 AAA domain of NVL2, which represents a novel mode of binding of a substrate with a type-II AAA-ATPase.

54

POSTER NUMBER: 2017

Architecture of a bacterial type II secretion system

Anastasia Chernyatina[1] and Harry H Low[1]*
[1]Department of Life Sciences, Imperial College, London, SW7 2AZ, UK
*Corresponding author

Bacterial type II secretion systems (T2SS) translocate virulence factors and enzymes across the cell envelope into the extra-cellular milieu. They are complex nanomachines that usually comprise 12 proteins from a single operon. The molecular mechanism of the T2SS has proven difficult to dissect in the absence of structural data for the entire machinery. Here we use a fusion of cryo and negative stain electron microscopy to reconstruct the T2SS from the human pathogen *Klebsiella pneumoniae*. We show that 7 proteins assemble in a stoichiometric relationship to form a ~2.5 MDa complex that spans the entire cell envelope. The outer membrane complex (OMC) comprises the secretin, the pilotin PulS and the GspC HR domain. It connects via long linkers to an energized inner membrane complex (IMC) where the hexameric ATPase GspE forms a central hub decorated by six highly flexible lobes. Our results show the architecture of the T2SS, and consequently the mechanism, to be different to other known bacterial secretion systems.

55

POSTER NUMBER: 2018

An integrated structural characterization of P-Rex1 activity in cancer cell signaling

Christina M Lucato[1,2], Michelle L Halls[3], Christina A Mitchell[1], James C Whisstock[1,2] and Andrew M Ellisdon[1,2]
[1]Department of Biochemistry and Molecular Biology, Monash University; [2]ARC Centre of Excellence in Advanced Molecular Imaging, Monash University; [3]Drug Discovery Biology Theme, Monash Institute of Pharmaceutical Sciences

P-Rex1 is a guanine nucleotide exchange factor that activates a number of Rho family GTPases that are integral in regulating cell growth and motility. Recently, P-Rex1 has been implicated in a number of cancer signaling pathways; its expression was shown to be necessary for melanoma dissemination and has been associated with increased metastatic phenotypes in both breast and prostate cancer. P-Rex1 is therefore a critical signal integrator in metastasis and abnormal cell growth, however currently there is very little mechanistic understanding of its activation and regulation. Crystallization and structural analysis of the catalytic DH-PH domains of P-Rex1 have allowed the elucidation of its mechanism of nucleotide exchange. However, DH-PH domain activity is inhibited by interactions with P-Rex1 C-terminal domains and these interactions have been mapped with cross-linking and mass spectrometry. Interestingly, these intramolecular interactions change significantly upon effector binding and understanding these changes has allowed for characterization of the multi-step P-Rex1 activation pathway. Further, molecular differences between autoinhibited P-Rex1 (185kDa) and effector-bound, partially activated P-Rex1 complexes (~230kDa) have been visualized with cryo-electron microscopy.

56

POSTER NUMBER: 2019

Structure of the mLST8-CCT Folding Complex

Jorge Cuéllar[1], Takuma Aoba[2], W. Grant Ludlam[2], Madhura Dhavale[2], Nicole C. Tensmeyer[2] Rebecca L. Plimpton[2], Aman Makaju[3], Sarah Franklin[3], Barry M. Willardson[2], and José M. Valpuesta[1]
[1]Centro Nacional de Biotecnología, Campus de la Universidad Autónoma de Madrid, 28049 Madrid, Spain, [2]Department of Chemistry and Biochemistry, Brigham Young University, Provo, UT 84602, and [3]Department of Internal Medicine, Nora Eccles Harrison Cardiovascular Research and Training Institute, University of Utah, Salt Lake City, UT 84112

The mechanistic target of rapamycin (mTOR) kinase is a master regulator of eukaryotic cell metabolism, growth and survival. mTOR functions in two distinct complexes, called mTORC1 and mTORC2, with mTORC1 regulating metabolic and growth pathways and mTORC2 contributing to cell survival pathways. While mTORC functions have been studied extensively, little is known about how the individual subunits of these complexes are assembled from their nascent polypeptide chains. Here, we provide evidence that an important chaperone for cytosolic proteins, the chaperonin containing TCP-1 (CCT), is involved in the folding and assembly of the -propeller protein mLST8 (mammalian lethal with SEC13 protein 8), a key component of both mTOR complexes. We have isolated an mLST8 folding intermediate bound to CCT and determined the structure of this complex to 4.4 Å resolution by single particle cryo-electron microscopy. This structure represents the highest resolution of a folding substrate inside the CCT cavity to date and shows mLST8 in an unexpected position between the two CCT octamer rings in a near-native conformation. Structures of other CCT substrates have positioned them within the folding cavity of one ring, but never between the rings. For example, the G protein ß subunit (G), which is a structurally similar -propeller protein to mLST8, sits near the upper rim of the folding cavity on one CCT ring. This unanticipated diversity of locations inside the CCT cavity suggests that the folding of mLST8, G and other substrates could be specifically targeted for molecular therapies that modulate their binding to CCT.

57

POSTER NUMBER: 1039

Purification of vaccinia viral a26 protein for biochemical analysis and crystallization

Yu-Chun Luo[1], Hao-Ching Wang[2], Wen Chang[1]
[1]Institute of Molecular Biology, Academia Sinica; [2]Graduate Institute of Translational Medicine, College of Medical Science and Technology, Taipei Medical University, Taipei, Taiwan

Vaccinia virus is the prototypic member of the *Orthopoxvirus* genus of the family *Poxviridae*. It contains a double-stranded DNA genome of approximately 190 kb and encodes more than 200 viral proteins. Our previous studies showed that vaccinia mature virus (MV) of Western Reserve (WR) strain is endocytosed into HeLa cells and subsequently viral membrane and vesicular membrane is fused to release viral core into cytoplasm. Furthermore, when a viral A26 protein was deleted from vaccinia virus genome, the resulting WR ΔA26 virus altered the entry route from endocytosis to plasma membrane fusion suggesting that removal of A26 protein eliminated virus dependence on acidification for membrane fusion. Further experiments showed that viral A26 binds to viral entry fusion complex in an acid-dependent manner and supported a model that A26 is a low-pH sensitive fusion suppressor during MV endocytosis into HeLa cells. Our current studies showed that A26 protein contains an acid sensitive N-terminal region that responds to low pH rendering virus-mediated fusion under control of environmental cue. When the N-terminal region of A26 was deleted or mutated vaccinia virus became unresponsive to low pH and mature virus infectivity was reduced more than 10 fold. To understand how A26 protein functions as an acid-sensitive fusion suppressor of vaccinia virus, we plan to express vaccinia viral A26 protein in *E. coli*, purify and crystallize the protein to obtain 3D structure in order to understand its structure and functional relationship.

58

POSTER NUMBER: 2021

Application of Parallel Continuous Simulated Tempering to Cryo-EM structural refinement at near-atomic resolution

Fengyun Ni[1], Qinghua Wang[1] and Jianpeng Ma[1,2]
[1]Verna and Marrs McLean Department of Biochemistry and Molecular Biology, Baylor College of Medicine, One Baylor Plaza, Houston, TX 77030, USA; [2]Department of Bioengineering, Rice University, 6500 Main Street, Houston, TX 77030, USA

Cryo-electron microscopy (Cryo-EM) has emerged as a promising technique for understanding molecular mechanisms of fundamental biological processes. Hardware advances in recent years have allowed near atomic to atomic-resolution single-particle Cryo-EM reconstruction for many macromolecular assemblies. However, model building and structural refinement at near atomic resolution remains a challenging task in the field. Our lab recently developed a powerful Parallel Continuous Simulated Tempering (PCST) algorithm that is highly efficient in accelerating barrier crossing and finding native structures than the conventional simulated annealing method. In this study, we used the PCST algorithm to refine Cryo-EM structures at near-atomic resolution in three systems at 6.8~8.0 Å resolution. Our results suggest that PCST-assisted Cryo-EM structural refinement not only significantly improved geometry statistics in terms of favored Ramachandran and rotamer distribution, but also enhanced the effective resolution of the resultant structural model by revealing many more native contacts that are only seen in atomic-resolution structures. In summary, PCST-assisted refinement represents an invaluable tool for Cryo-EM structural refinement.

59

POSTER NUMBER: 2022

Cryo-EM study of RISC loading complex Dicer2-R2D2

Shichen Su[1], Jia Wang[2], Xun Yuan[3], Ying Huang[3*], Hongwei Wang[2*], Jinbiao Ma[1*],
[1]State Key Laboratory of Genetic Engineering, Collaborative Innovation Centre of Genetics and Development, Department of Biochemistry, Institute of Plant Biology, School of Life Sciences, Fudan University, Shanghai, China; [2]Ministry of Education Key Laboratory of Protein Sciences, Beijing Advanced Innovation Center for Structural Biology, School of Life Sciences, Tsinghua University, Beijing 100084, China; [3]State Key Laboratory of Molecular Biology, National Center for Protein Science Shanghai, Shanghai Science Research Center, Shanghai Key Laboratory of Molecular Andrology, CAS Center for Excellence in Molecular Cell Science, Shanghai Institute of Biochemistry and Cell Biology, Chinese Academy of Sciences; University of Chinese Academy of Sciences, China
*Co-Corresponding Authors

RISC loading complex (RLC) is required for small interference RNAs (siRNAs) assembly into the RNA-induced silencing complex (RISC). In Drosophila melanogaster, RLC contains the Dicer2-R2D2 complex that binds siRNA in an asymmetric manner based on thermodynamic stability and recruits it to Argonaute2 (Ago2). The asymmetric binding manner of siRNA by RLC determines which strand of siRNA loading into RISC. However, the molecular mechanism of RLC binding siRNA remain unclear. We are using cryo-EM and single-particle analysis of Dicer2-R2D2 complex purified from insect cell to gain insight into the structural basis for substrate preference of Drosophila RLC. Most recent progress of this work will be reported in the meeting.

60

POSTER NUMBER: 2023

Using cryo-electron tomography and sub-tomogram averaging to study the integrated virus fusion machinery in intact influenza virions

Vidya Mangala Prasad*, Nancy Hom* and Kelly K. Lee
University of Washington, USA

Influenza A virus is a major zoonotic and human pathogen that exhibits significant antigenic and phenotypic variability. Although influenza viruses have been extensively studied, we have a limited arsenal of antiviral agents and a critical need for improved vaccines that can protect against infection. Here, we seek to study the integrated influenza virus fusion machinery, namely the hemagglutinin (HA) fusion protein and the matrix (M1) protein in intact virions, under near physiological states using cryo-electron tomography (cryo-ET) and sub-tomogram averaging. This enables us to understand their critical roles in membrane fusion, which ensures delivery of the viral genome into host cells.

Although detailed, static structures of the HA fusion protein's ectodomain are known, mechanisms of the dramatic reorganization of HA that drives membrane fusion are still being unraveled. On the other end of the spectrum, the M1 protein remains poorly characterized, despite its pervasive influence on many steps in the virus infectious cycle. The M1's structure, organization, and interactions with other virus components still remain to be determined. HA and M1 act in concert during virus fusion, but their sites and extent of interactions have evaded direct characterization.

Using cryo-ET, we are delineating the structure and conformations of these essential components in the filamentous Udorn influenza virus strain. The ~220 kDa HA trimers are densely packed on the virus surface. The native structure of HA, including transmembrane domains, has been determined using sub-tomogram averaging to ~15Å resolution; we anticipate being able to reach sub-nanometer resolution. Thousands of M1 subunits form a helical matrix array that lines the inner leaflet of the viral membrane. Using sub-tomogram averaging, the structure of M1 (~28kDa) in the helical array is beginning to be resolved. These studies will allow us to identify regions of contact between the HA, the viral membrane and the M1, in the authentic infectious virion. This will shed light on critical interactions that enable these proteins to act synergistically as they facilitate membrane fusion and delivery of ribonucleoprotein gene segments into host cells.

Funding: National Institute of Health (R01-GM099989)

61

POSTER NUMBER: 2024

Anti-diabetic drug binding site in KATP channels revealed by Cryo-EM

Greogry M Martin[1], Balamurugan Kandasamy[1], Frank DiMaio[2], Craig Yoshioka[3]*, & Show-Ling Shyng[1]*
[1]Department of Biochemistry and Molecular Biology, Oregon Health & Science University, Portland, OR 97239, USA; [2]Department of Biochemistry, University of Washington, Seattle WA 98195, USA; [3]Department of Biomedical Engineering, Oregon Health & Science University, 2730 SW Moody Ave, Portland, OR 97239, USA
*Corresponding Author

Sulfonylureas are widely-prescribed anti-diabetic medications that act by inhibiting pancreatic K_{ATP} channels, large hetero-octameric complexes comprising an ABC transporter SUR1, and an inward-rectifier potassium channel Kir6.2. The mechanism by which these drugs interact with and inhibit the channel has been extensively investigated, yet it remains unclear where the drug binding pocket resides. Here, we present a cryo-EM structure of the channel bound to a high-affinity sulfonylurea drug glibenclamide and ATP at 3.7Å resolution, which reveals unprecedented details of the ATP and glibenclamide binding sites. Importantly, the structure shows for the first time that glibenclamide is lodged in the transmembrane bundle of the SUR1-ABC core connected to the first nucleotide binding domain near the inner leaflet of the lipid bilayer. Mutation of residues predicted to interact with glibenclamide in our model led to reduced sensitivity to glibenclamide. Our structure provides novel mechanistic insights of how sulfonylureas and ATP interact with the K_{ATP} channel complex to inhibit channel activity.

This work was supported by the National Institutes of Health grants R01DK066485 (to S.-L. S.) and F31DK105800 (to G.M.M.)

62

POSTER NUMBER: 2025

Conserved structural motifs can be used to improve cryoEM modeling of voltage gated ion channels

Wilnelly Martinez-Ortiz and Timothy J. Cardozo*
Department of Biochemistry and Molecular Pharmacology, NYU School of Medicine, NY, NY
*Corresponding Author

The super resolution revolution has recently and rapidly catapulted cryo-electron microscopy (cryoEM) to the forefront of state-of-the-art structure determination techniques, given its capacity to routinely achieve atomic or near atomic 3D structures of complex macromolecules such as voltage-gated ion channels (VGIC). Recently, several VGIC's have been resolved at near atomic resolution by single-particle non-crystalline cryoEM. Despite the numerous tools available to determine the relative resolution and overall model fitness to raw experimental data, to-date, there are no tools capable of assessing the quality of the residue assignment in the resultant structural models. To address this void, and in an effort to assess the accuracy of all VGIC structures deposited in the Protein Data Bank (PDB), we developed a bioinformatics method that scores 3D models of the pore-forming subunit of the VGIC's based on the preservation of evolutionarily conserved functional contacts within the voltage-sensing-domain (VSD) structural fold. We found that our scoring function is capable of rapidly assessing the accuracy of all reported VGIC structures independent of the source of the raw data, serving as an accuracy metric that can be used to compare cryo-EM and X-ray crystal structures. Importantly, any subtle molecular modeling errors in VGIC-VSD models is unambiguously revealed with our metric. Surprisingly, several of these recent single-particle VGIC cryoEM structural models received low scores for accuracy due to simple residue assignment errors in their VSDs. Using constraints derived from our metric in modeling the structures from the experimental map revealed that with our method near-atomic cryoEM data can be easily refined to 3D models with accuracy equivalent to atomic resolution.

Acknowledgements: NIH grants: F31HL124898-01-WMO and DP2 OD004631-TC

63

POSTER NUMBER: 2026

Global conformational changes in RNA polymerase accompany nascent RNA hairpin-stabilized transcriptional pause

Tatiana V. Mishanina[1], Jin Young Kang[2], Rachel A. Mooney[1], Seth A. Darst[2], and Robert Landick[1]
[1]Department of Biochemistry, University of Wisconsin, Madison, WI USA
[2]The Rockefeller University, 1230 York Avenue, New York, NY, USA

Cellular transcription of DNA-encoded information into RNA is a complex, highly regulated process carried out by multi-subunit RNA polymerases (RNAPs). Transcription by RNAPs is not a continuous process, but is instead interrupted by DNA sequence-dependent pauses, during which the polymerase remains bound to the nucleic acids without active RNA synthesis occurring. Transcriptional pausing plays crucial regulatory roles in both prokaryotes and eukaryotes. Pausing in bacteria allows synchronization of transcription and translation, creates time windows for transcription regulators to bind the elongating transcription complex, serves as an initial step in transcription termination, and promotes the proper folding of nascent RNA transcripts. Some nascent RNA structures thought to form *en route* out of RNAP prolong transcriptional pauses ten-fold or more. Although important, the underlying mechanism of such pauses remains poorly understood. Here we sought to gain insight into the pause mechanism by determining a 3.8 Å cryo-EM structure of a paused transcription complex stabilized by an RNA hairpin. The structure revealed a series of features unique to the paused complex, including the presence of a double-stranded RNA hairpin; a rearrangement of several large modules of RNAP, and an incomplete movement of the nucleic acids that would preclude addition of the next nucleotide. We tested some of these features biochemically by installing a series of crosslinks in RNAP and measuring the effect on pausing. Altogether, our results suggest a molecular mechanism where concerted conformational changes in the transcription complex, assisted by the nascent RNA structure, inhibit further RNA synthesis.

64

POSTER NUMBER: 2027

HIV-1 Nef as a cargo-selective switch in hijacking AP-1 clathrin adaptor complexes

Kyle L. Morris[1], Cosmo Buffalo[1], Xuefeng Ren[1], and James H. Hurley[1]
[1]Department of Molecular and Cell Biology and California Institute for Quantitative Biosciences, University of California, Berkeley, Berkeley, CA 94720, USA

The lentiviral protein Nef promotes infectivity and subverts immune surveillance of infected cells by downregulating cell surface proteins such as SERINC5 and MHC-I. The Nef proteins of HIV-1 group O, HIV-2, and SIV also target the cell surface restriction factor tetherin. These downregulatory activities are mediated by Nef hijacking the clathrin adaptor proteins AP-1 and AP-2 so as to redirect cargoes away from the cell surface. Nef hijacks AP-1 in part by enhancing its Arf1-dependent trimerization. Nef and Arf1 in combination promote structurally and functionally distinct closed (trapped) and open (active) trimeric conformations. A new 3.8 Å cryo-EM structure of the AP-1:Arf-1:NL4-3 Nef:tetherin closed trimer reveals novel contacts that stabilize this hijacked complex. These data are interpreted in terms of a mechanism for trapping non-preferred cargoes such as tetherin such that they cannot function at the cell surface.

65

POSTER NUMBER: 2028

***Corpora amylacea* are dense membranous labyrinths containing intact mitochondria and vesicles**

Paula P. Navarro[1], Christel Genoud[2], Daniel Castaño Díez[1], Alexandra Graff-Meyer[2], Yvonne de Gier[3], Matthias Lauer[4], Markus Britschgi[5], Bernd Bohrmann[5], Stephan Frank[6], Annemieke J.M. Rozemuller[7], Wilma D.J. van de Berg[3], Henning Stahlberg[1*], Sarah H. Shahmoradian[1*],
[1]Center for Cellular Imaging and NanoAnalytics (C-CINA), Biozentrum, University of Basel, Basel, Switzerland; [2]Friedrich-Miescher Institute for Biomedical Research, Basel, Switzerland; [3]Department of Anatomy and Neurosciences, Section: Clinical Neuroanatomy, AO|2M, Amsterdam Neuroscience, VU University Medical Center, Amsterdam, The Netherlands; [4]Roche Pharma Research and Early Development, Chemical Biology, Roche Innovation Center Basel, Basel, Switzerland; [5]Roche Pharma Research and Early Development, NORD DTA/Neuroscience Discovery, Roche Innovation Center Basel, Basel, Switzerland; [6]Division of Neuropathology, Institute for Pathology, University Hospital Basel, Basel Switzerland; [7]Department of Pathology, Amsterdam Neuroscience, VU University Medical Center (VUmc), Amsterdam, The Netherlands
*Corresponding Author

Corpora amylacea are cell-derived aggregates that appear prominently in the aging human brain. While their histological identification is straightforward, their ultrastructural composition and microenvironment at the nanoscale is unclear. Here, we apply correlative serial block-face scanning electron microscopy and transmission electron tomography to gain three-dimensional insight on the ultrastructure and surrounding microenvironment of *Corpora amylacea* in aged human brain tissue samples. We found that they are composed of dense labyrinth-like sheets of lipid membranes, contain vesicles and morphologically intact mitochondria, and are in close proximity to blood vessels and the glymphatic system of the brain, primarily within astroglia. Our results clarify the nature of such aggregates and lead to a better understanding of how they may arise and develop in the aging brain.

66

POSTER NUMBER: 2029

Cryo-EM structure of human telomerase and new insights into its assembly and function

Thi Hoang Duong Nguyen[1,3], Jane Tam[1], Robert A. Wu[1], Basil J. Greber[1,3], Kathleen Collins[1*], Eva Nogales[1,2,3*]
[1]University of California, Berkeley; [2]Howard Hughes Medical Institute; [3]Lawrence Berkeley National Laboratory
*Corresponding authors

Due to the end-replication problem, telomeric caps at the ends of eukaryotic linear chromosomes are progressively shortened, which eventually leads to genome instability. To maintain genome integrity, the telomerase ribonucleoprotein (RNP) synthesizes telomeric repeats through its telomerase reverse transcriptase (TERT) and an integral RNA subunit (hTR) carrying the template for repeat synthesis[1]. In addition to TERT and hTR, the human telomerase holoenzyme consists of a number of other protein factors required for RNP assembly and localization. Telomerase has been shown to play important roles in cancer, aging and a number of human diseases. Due to its cellular scarcity and complex RNP assembly, the composition of human telomerase is still a matter of debate and available structural information on human telomerase is limited to a 25Å negative stain reconstruction[2]. Here we present biochemical and cryo-EM studies on the human telomerase holoenzyme. Our results reveal, for the first time, the detailed architecture of an active telomerase RNP from a multicellular organism, giving new insights into its assembly and function.

References:
[1]Blackburn, E. H. & Collins, K. Telomerase: an RNP enzyme synthesizes DNA. *Cold Spring Harb. Perspect. Biol.* **3**, a003558 (2011).
[2]Sauerwald, A. et al. Structure of active dimeric human telomerase. *Nat. Struct. Mol. Biol.* 20, 454-461 (2013).

67

POSTER NUMBER: 2031

Conformational Plasticity in Viral Glycoproteins

Jesper Pallesen
The Scripps Institute, San Deigo, CA, USA

Enveloped viruses carry fusion glycoproteins on their envelope surfaces. Fusion of viral envelopes and host cell membranes is catalyzed by these fusion glycoproteins and mark the principal onset of infection. Fusion glycoproteins are complex molecular machines that undergo massive conformational transformation from a metastable prefusion state to an energetically more favorable postfusion state when catalyzing their primary function. Fusion triggering mechanisms can be any combination of receptor recognition, enzyme cleavage or lowering of pH. Little is known about the conformational intermediates that are involved in the work path of fusion glycoproteins but evidence from multiple experimental approaches is accumulating that multiple conformations are probed prior to fusion. Employing sophisticated classification protocols on cryo-EM data of conformationally heterogeneous immune complexes, we report the existence of glycoprotein prefusion conformational plasticity across multiple virus families and we report a fusion intermediate of the receptor-bound HIV-1 Env fusion glycoprotein.

68

POSTER NUMBER: 2032

Mechanistic insights into PIKK family function: High resolution cryoEM studies of ATM, ATR and DNA-PK provide a structural basis to understanding the regulation of the DNA damage response

Chris Phillips[1], Domagoj Baretic[2], Taiana Maia De Olivera[1], Paul Wan[1], Hannah Pollard[1] David Fisher[1], Christopher M. Johnson[2], Balaji Santhanam[2] Caroline Truman[1], Sjors Scheres[2] and Roger L. Williams[2]
[1]Discovery Sciences, IMED Biotech Unit, AstraZeneca, Cambridge, UK
[2]MRC Laboratory of Molecular Biophysics, Cambridge, UK

Three members of the phosphatidylinositol 3-kinase-related kinase (PIKK) family of kinases play prominent roles in the cellular response to DNA damage; ATM, ATR and DNA-PK. Ataxia telangiectasia mutant (ATM) is the key regulatory trigger for the repair of DNA double strand breaks (DSBs), a taxia telangiectasia and Rad3-related protein (ATR) is activated mainly by single-stranded DNA (ssDNA) ends that are generated following the induction of DNA adducts or during the processing of DSBs, while DNA-PK is a major player in the repair of DSBs induced by genotoxic stresses or during V(D)J recombination, via non-homologous end-joining (NHEJ). Together these three large kinases lie at the core of DNA damage response (DDR), a complex network of interacting pathways that lead either to damage repair or to programmed cell death. DDR is a highly promising area for cancer drug discovery as inhibition of the pathways exploits the intrinsic genomic instability of cancer cells and inhibitors of each these three kinase are currently in clinical development.

Here we discuss novel insights gained from high resolution cryoEM structures for each of these core DDR regulators.

69

POSTER NUMBER: 3001

Structure of the Essential Eukaryotic Grc3/Las1 RNA Processing Machinery

Monica C. Pillon[1], Mack Sobhany[1], Mario J. Borgnia[2], Jason G. Williams[3], and Robin E. Stanley[1]
[1]Signal Transduction Laboratory, National Institute of Environmental Health Sciences, National Institutes of Health, Department of Health and Human Services, 111 T. W. Alexander Drive, Research Triangle Park, NC 27709, USA; [2]Laboratory of Cell Biology, Center for Cancer Research, National Cancer Institute, National Institutes of Health, Bethesda, MD 20892, USA; [3]Epigenetics and Stem Cell Biology Laboratory, National Institute of Environmental Health Sciences, National Institutes of Health, Department of Health and Human Services, 111 T. W. Alexander Drive, Research Triangle Park, NC 27709, USA

Las1 is a recently discovered endoribonuclease that collaborates with Grc3-Rat1-Rai1 to process precursor ribosomal RNA (pre-rRNA), yet its mechanism of action remains unknown. Disruption of the mammalian Las1 gene has been linked to congenital lethal motor neuron disease and X-linked intellectual disability disorders, thus highlighting the necessity to understand Las1 regulation and function. Here we report that the essential Las1 endoribonuclease displays weak RNase activity, however in the presence of the polynucleotide kinase Grc3, Las1 is reprogrammed to efficiently and specifically cleave RNA. Our results establish that Grc3 specifically directs Las1 endoribonuclease cleavage to its targeted C2 site both *in vitro* and in *Saccharomyces cerevisiae*. Moreover, we observed Las1-dependent activation of the Grc3 kinase activity exclusively towards single-stranded RNA. Using negative stain electron microscopy, we reveal that Las1 and Grc3 assemble into a tetrameric complex that is required for competent rRNA processing. The tetrameric Grc3/Las1 crosstalk draws unexpected parallels to endoribonucleases RNaseL and Ire1, and establishes Grc3/Las1 as a novel member of the RNaseL/Ire1 RNA Splicing Family. Together, our work provides mechanistic insight for the regulation of the Las1 endoribonuclease and identifies the tetrameric Grc3/Las1 complex as a new example of a protein guided programmable endoribonuclease.

70

POSTER NUMBER: 2038

Molstack - platform for comparative presentation of density maps and their interpretations

Przemyslaw J. Porebski, Piotr Sroka, Grzegorz Bokota, Wladek Minor
Department of Biological Physics & Molecular Physiology, University of Virginia, Charlottesville, Virginia

The models of the biomolecular structures are predominantly based on the interpretation of the underlying experimental densities. These models are partially subjective therefore should always be inspected and should be able to stand against alternative interpretations. To facilitate these tasks we have created Molstack - a web-tool

that organizes the sharing and dissemination of these structural models along with their experimental evidence (density) as an interactive session.
Initially developed for presentation of electron density maps from X-ray crystallography Molstack has been extended to handle other types of densities. Molstack allows researchers to present the evidence of their interpretation and reviewers and readers can independently judge the experimental evidence of the authors' conclusions. The synchronized side-by-side views allow to present and compare combinations of different maps and models providing the researchers with a tool to present, privately share or even publish their new hypotheses in the context of prior results, alternate models, and multiple interpretations. It makes it also an useful tool for presenting maps resulting from different types of data processing, map calculation protocols and map post-processing as well as more speculative models that would not be otherwise publishable.
The server is available at http://molstack.bioreproducibility.org.

71

POSTER NUMBER: 2039

An update on detergent usage in Cryo-EM structure determination of membrane proteins

Edward Pryor[1], Ben Travis[1]
[1]Anatrace Products, LLC, Maumee, OH 43537

The past few years have been revolutionary for the field of single-particle electron cryo-microscopy (cryo-EM), with over 50% of the total deposited structures being determined since 2014. Currently, there are over 950 unique (<95% sequence identity) cryo-EM structures deposited in the PDB, 80 of which are membrane proteins. Although we are witnessing many significant strides in this field, challenges in membrane protein production, purification, and structure determination still persist. These challenges often revolve around the choice of detergent used to solubilize and purify the membrane protein, as it is well established that the detergent used can have drastic effects on the stability and activity of the protein. Thus, choosing the correct detergent is of paramount importance.

We have built a database containing detergent usage data for all unique membrane protein structures deposited in the PDB. Here, we'll present data on which detergents are commonly used in the Cryo-EM structure determination of membrane proteins, and examine if any trends in detergent usage exist among specific classes of membrane proteins. The results from this analysis show that a large number of detergents are compatible and frequently used with Cryo-EM, including DDM, LMNG, and Amphipol A8-35. Additionally, we will highlight examples where the use of fluorinated surfactants, such as Octyl Maltoside and Fos-Choline-8, have been used to improve the vitrification process.

72

POSTER NUMBER: 3002

Hand-over-hand: How ATP drives protein translocation

Cristina Puchades[1,2], Anthony J. Rampello[3], Mia Shin[1,2], Christopher J. Giuliano[3], R. Luke Wiseman[2], Steven E. Glynn[3]*, Gabriel C. Lander[1]*
[1]Department of Integrative Structural and Computational Biology, [2]Department of Molecular Medicine, The Scripps Research Institute, La Jolla CA 92037, [3]Department of Biochemistry and Cell Biology, Stony Brook University, Stony Brook, NY 11794
*Co-corresponding Authors

We present the first atomic model of a substrate-bound inner mitochondrial membrane AAA+ quality control protease, YME1. Our ~3.4 Å cryo-EM structure reveals how the ATPases form a closed spiral staircase encircling an unfolded substrate, directing it toward the flat, symmetric protease ring. Importantly, the structure reveals how three coexisting nucleotide states allosterically induce distinct positioning of tyrosines in the central channel, resulting in substrate engagement and translocation to the negatively charged proteolytic chamber. This tight coordination by a network of conserved residues defines a sequential, around-the-ring ATP hydrolysis cycle that results in step-wise substrate translocation. Furthermore, we identify a hinge-like linker that accommodates the large-scale nucleotide-driven motions of the ATPase spiral independently of the contiguous planar proteolytic base. These results define the first molecular mechanism for a mitochondrial inner membrane AAA+ protease and reveal a translocation mechanism likely conserved for other AAA+ ATPases.

Reference:

Atomic structure of the mitochondrial inner membrane AAA+ protease YME1 reveals the mechanism of substrate

processing. Puchades *et al.* Science (2017). Accepted.
Preprint version available in BioRxiv: https://doi.org/10.1101/189316

73

POSTER NUMBER: 3003

Structure of a histone H3K4 methyltransferase COMPASS

Qianhui Qu[1], Yohhei Takahashi[3], Yidai Yang[4], Hongli Hu[1], Yan Zhang[1], Jean-Francois Couture[4], Ali Shilatifard[3], Georgios Skiniotis[1,2]
[1]Department of Molecular and Cellular Physiology, [2]Department of Structural Biology, Stanford University; [3]Department of Biochemistry and Molecular Genetics, Feinberg School of Medicine, Northwestern University; [4]Department of Biochemistry, Microbiology, and Immunology, Ottawa Institute of Systems Biology, University of Ottawa, Canada

Methylation of nucleosomal histone H3K4, generally associated with active gene transcription in all eukaryotes, is mainly mediated by the evolutionarily conserved MLL/COMPASS-like (complex associated with Set1 protein) complexes. Aberrant activity or genetic rearrangement of MLL/COMPASS-like complexes has been increasingly linked to human diseases, especially leukemia. The activity of the methyltransferase Set1/MLL protein is highly regulated through its incorporation into a macromolecular complex that includes a common set of conserved proteins. Specifically, Cps30/WDR5, Cps50/RbBP5, Cps60/ASH2L and Cps25/DPY30 are shared among all MLL/COMPASS-like complexes from yeast to humans, respectively. Here we employed cryo-EM to determine the structure of the yeast Set1/COMPASS catalytic core complex including Cps25/30/40/50/60 and Set1 proteins. The structure shows that Cps50 serves as an assembly hub by interacting with most subunits of the core COMPASS complex. The catalytic SET domain is positioned at the junction of the Y-shaped assembly and is extensively contacted by its core partners. Of note, the predicted mobile SET-I motif, whose relative position is sensitive for methylation activity, is highly coordinated by Cps30, Cps50 and Cps60. This structure provides the framework to understand the regulation of methyltransferase activity within the MLL/COMPASS-like family.

74

POSTER NUMBER: 3005

Ribosome biogenesis: emerging view for a central role of assembly factors in the maturation of the ribosomal subunits functional core using Cryo-EM

Aida Razi and Joaquin Ortega
Department of Anatomy and Cell Biology, McGill University, Montreal, Quebec, Canada

The ribosome is the enzyme responsible for protein synthesis and the most complex macromolecule existing in bacterial cells. In *Escherichia coli* the ribosome is made from more than 50 different components organized into the small (30S) and large (50S) ribosomal subunits (1). Understanding how the components of the bacterial ribosome come together and organize themselves still remains a daunting challenge. Despite its complexity, each bacterial cell assembles more than 20,000 ribosomes in less than 30 minutes. Cells are able to maintain this assembly rate because a number of auxiliary factors make the process extremely efficient (2). Our work focuses on the YjeQ assembly factor. Recent work indicates that this protein assists the late stages of assembly of the 30S subunit (3). To understand the precise mechanisms on how YjeQ performs its function, we have obtained the structure of YjeQ in complex with the 30S subunit at ~5 Å resolution using cryo-electron microscopy (4). The cryo-EM map shows the highly conserved Phe48 residue of YjeQ disturbs helix 44, an essential ribosomal motif for the decoding process. Analysis of this region revealed that A1492 flips out from the helix. This conformation resembles that exhibited by nucleotides A1492-A1493 during proofreading of the codon-anticodon pairing. This finding suggests that YjeQ has a role as a checkpoint protein dedicated to test the decoding fidelity of the 30S subunit before the particle is released to the pool of actively translating ribosomes. This finding constitutes the first example of a bacterial ribosome assembly factor with a checkpoint role.

1. Chen et al. (2013) *J Mol Biol* 425(4):767-779.
2. Razi et al. (2016) *Nucleic Acids Res.* 45(3):1027-1040.
3. Jeganathan et al. (2015) *RNA* 21(6):1203-1216.
4. Razi et al. (2017) *Proc Natl Acad Sci U S A* 114(17):E3396-E3403.

75

POSTER NUMBER: 3006

Structural and Functional Characterization of a Cyclic Nucleotide-gated Ion Channel

Jan Rheinberger[1], Philipp Schmidpeter[1], Xiaolong Gao[1], Crina Nimigean[1*]
[1]Departments of Anesthesiology, Weill Cornell Medicine, New York, NY, 10065, United States.
*Corresponding author

Cyclic nucleotide-modulated ion channels, namely cyclic nucleotide-gated (CNG) and hyperpolarization-gated cyclic nucleotide modulated (HCN) ion channels, play essential roles in various physiological processes, such as, photon detection in photoreceptors, odorant sensation by olfactory neurons and pace making activity in the heart and brain. Previous studies provided insights into structure and function of these channels, but little is known about the function of the purified channel proteins in a defined environment and how the conformations available from structural studies are related to functional states. To address these questions, we established expression and purification of a cyclic nucleotide-gated ion channel as a model system. A combination of structural determination by single particle cryo-electron microscopy and function characterization using bilayer electrophysiology and stopped-flow Tl+ flux assay measurements was used to reveal the characteristics of purified channels in a lipid environment. The channel is activated by cyclic nucleotides and voltage has a major impact on its open probability. Channel structures in different ligand-bound states allowed us to determine key regions for its function. Our structural and functional characterization shows that our channel is a good model system to understand gating and ligand modulation in cyclic nucleotide-modulated channels.

76

POSTER NUMBER: 3007

CaBLAM - A New Tool Designed to Validate 2.5-4Å Models from CryoEM

Jane S. Richardson, Christopher J. Williams, David C Richardson
Duke University, Durham NC 27710

Many of the basic model validation tools from x-ray crystallography lose their diagnostic power and even become misleading at the 2.5-4Å resolution range now being pioneered by the new cryoEM methods. This is largely because they analyze phi,psi or H-bonding, which depend on the backbone carbonyl O orientations that are poorly seen or not at all in this range, especially for cryoEM. This new need has prompted development of two quite novel tools: EMringer from UCSF and CaBLAM from our lab at Duke.

The new idea in CaBLAM is using Calpha virtual dihedrals, the most robustly observed and interpreted feature at these resolutions, to define the probable local backbone trace, then adding (in 3D) the virtual dihedrals between successive COs to diagnose where the model is wrong and, often, how it can be fixed. The current implementation in Phenix and MolProbity is good at identifying helices and strands even when they are disguised by highly distorted peptide orientations. CaBLAM can be useful in high-resolution cryoEM for three kinds of tasks: 1) to make individual structure models more accurate; 2) to validate relative model accuracy between structures; and 3) to speed and evaluate the development of automated software for building these models. Examples will be shown from assessment in the recent CryoEM Model Challenge.

77

POSTER NUMBER: 3008

Inhibitory Role of Functionalized Fullerenes in HIV-1 Maturation

Alejandro Rodriguez[1], Zachary S. Martinez[1], Edison Castro[1], Luis Echegoyen[1*], Manuel Llano[1*], and Ricardo A. Bernal[1*]
[1]University of Texas at El Paso
*Corresponding Author

HIV-1 is the most prevalent of the major HIV strains currently affecting the human population. Several anti-HIV drugs exist, but drug resistant strains emerge rapidly, rendering treatments useless in many cases. For decades, it has been known that fullerene compounds inhibit the HIV-1 life cycle, but the mechanism of this has not been elucidated. Molecular docking simulations proposed that fullerene compounds interact with the HIV-1 protease, inactivating it and preventing the processing of the Gag polyprotein. Recent studies have invalidated this claim and have demonstrated that functionalized C_{60} and C_{70} fullerene compounds prevent the maturation of HIV-1 in a protease-independent manner. In this work we attempt to elucidate the interaction between derivatized fullerene compounds and an HIV-1 target protein in-vitro. Key aspects of this work include the cloning, expression, purification, and visualization of the proteins using electron microscopy, along with the assays that will highlight the mechanism by which fullerene compounds inhibit the maturation of HIV-1. These findings can help pinpoint novel target sites in HIV-1 proteins, opening the way for the production of new drugs.

78

POSTER NUMBER: 3009

The 3.5 Å CryoEM Structure of Nanodisc Reconstituted Yeast Vacuolar ATPase Vo Proton Channel

Soung-Hun Roh[1,2,4], Nicholas J. Stam[3], Corey F. Hryc[4], Sergio Couoh-Cardel[3], Grigore Pintilie[4], Wah Chiu[1,2,4] and Stephan Wilkens[3]
[1]Department of Bioengineering and Bio-X Center, Stanford University, Stanford; [2]Biosciences Division, SLAC National Accelerator Laboratory; [3]Department of Biochemistry and Molecular Biology, SUNY Upstate Medical University; [4]Verna and Marrs McLean Department of Biochemistry and Molecular Biology, Baylor College of Medicine

The molecular mechanism of transmembrane proton translocation in rotary motor ATPases is not fully understood. Here we report the 3.5 Å resolution cryoEM structure of the lipid nanodisc-reconstituted V_0 proton channel of the yeast vacuolar H^+-ATPase (V-ATPase), captured in a physiologically relevant, autoinhibited state. The resulting atomic model provides unprecedented detail for the amino acid residues at the interface of the proteolipid ring and the transmembrane portion of subunit *a* that constitute the proton pathway. Together with previous mutagenesis studies, we propose the chemical basis of transmembrane proton transport. Moreover, we discovered that the C-terminus of the assembly factor Voa1 is an integral component of mature V_0. Voa1's C-terminal transmembrane α helix is bound inside the proteolipid ring, where it contributes to the stability of the complex. Our structure rationalizes possible mechanisms by which mutations in human V_0 can result in disease phenotypes, and may thus provide new avenues for therapeutic interventions.

79

POSTER NUMBER: 2040

Mechanistic insights from the cryo-EM visualization of the microtubule severing enzyme katanin

Elena Zehr, Agnieszka Szyk, Greg Piszczek, Xiaobing Zuo and Antonina Roll-Mecak
Porter Neuroscience Research Center, National Institutes of Health

Microtubule-severing enzymes katanin, spastin and fidgetin are AAA ATPases important for the biogenesis and maintenance of complex microtubule arrays in axons, spindles and cilia. They generate internal breaks in microtubules thus modulating their dynamics and organization. They are found throughout the animal kingdom, plants and protozoa, and mutations in microtubule severing enzymes cause severe neurodegenerative and neurodevelopmental disorders. Because of a lack of 3D structures, the mechanism of microtubule severing remains poorly understood 25 years after its discovery. Here we report the first X-ray structure of the monomeric AAA katanin module and cryo-EM reconstructions of the hexamer in two conformations at 4.4Å and 6Å resolution. These reveal an unexpected asymmetric arrangement of the AAA domains mediated by structural elements unique to severing enzymes that are critical for their function. Our reconstructions show that katanin cycles between open spiral and closed ring conformations, depending on the ATP occupancy of a gating protomer that tenses or relaxes inter-protomer interfaces and closes a 40Å wide gate in the katanin hexamer. Cycling of the hexamer between these conformations would provide the power stroke for microtubule severing. Our structures illuminate the mechanism of action of katanin and spastin disease mutations and together with structure guided functional studies provide a structural framework for understanding microtubule severing

80

POSTER NUMBER: 3010

ModelMaker: An interactive integrative modeling tool to build complete macromolecular machines

Ryan McGreevy[1], Maximilian Scheurer[2], Klaus Schulten[1], Emad Tajkhorshid[1], Till Rudack[3,1,*]
[1]NIH Center for Macromolecular Modeling and Bioinformatics, University of Illinois at Urbana Champaign; [2]Biochemistry Center, University of Heidelberg; [3]Department of Biophysics, Ruhr University Bochum
*Corresponding Author

Recent revolutionary advances in cryo-electron microscopy (cryo-EM) provide an unprecedented opportunity to characterize at atomic detail the structure and function of the macromolecular machineries that are crucial to the function of living cells. However, deriving complete structural models from cryo-EM densities creates challenges, especially when modeling highly flexible or multi-conformational domains. To overcome these challenges, we developed an integrative modeling tool called ModelMaker, to interactively build complete models into cryo-EM densities (3-12 Å resolution) guided by incomplete structural data from experiments, automated structure prediction, and molecular dynamics flexible fitting (MDFF). ModelMaker is an all-in-one modeling solution combining the strength of established but often complicated to use modeling suits like Rosetta and Modeller with MDFF in an easy-to-use manner in VMD. Integrative modeling approaches usually aim to automate the process of structure analysis to avoid human bias, yet the experience of structural biologists may be a desirable factor in structure refinement. As a unique feature ModelMaker allows for incorporation of user expertise by taking advantage of interactive MDFF, allowing for manual manipulation of the structure, and therefore prevents fitting simulations from getting stuck in unreasonable minima. ModelMaker enabled us to construct the first high-resolution (3.9 Å) nucleotide-bound atomic structural model of the human 26S proteasome [1], a 2.5 MDa multi-subunit molecular machine, which is a key player for protein degradation in cells and an important drug target for multiple diseases. In a follow up study, we derived four more structural models of the yeast proteasome in different conformational states [2]. These models provide the first atomic insights as to how ATP hydrolysis in the engine of the proteasome unwinds proteins and steers them towards the degradation chamber.

References
[1] Schweitzer, Aufderheide, Rudack, et al (2016) PNAS 113:7816.
[2] Wehmer, Rudack, Beck et al. (2017) PNAS 114:1305.

81

POSTER NUMBER: 3011

Understanding AAV DNA packaging

Vishaka Santosh
Virginia Commonwealth University, Physiology and Biophysics, Richmond, VA, USA

Adeno-associated virus (AAV) has been utilized as a gene therapy vector, but the biology of the virus is unknown. One of the mysteries of AAV is how exactly the virus packages the genome. That being said, the non-structural Rep proteins appear to be necessary and sufficient for proper AAV packaging. The small Rep proteins serve as the molecular motor to package the DNA into the viral capsid, but the large Rep proteins' role has not been clarified. Furthermore, there is a consensus all Rep proteins should have the same helicase activity, since they have the same helicase domain, and therefore should serve the same role. That actually isn't the case since only Rep40 appears to be necessary for this task, so helicase assays were conducted to assess the activity of each of the Rep proteins. We found that Rep40 has a higher helicase activity compared to Rep68 implying its role as the molecular motor as logical. Since Rep68 doesn't unwind DNA nearly as well as Rep40, we hypothesized another role that is linked to its oligomeric capacity. Using AUC, we have discovered Rep68's ability to not only form higher order molecular complexes on its own, but depending on the DNA substrate, the molecular weight of the DNA/protein complex can change. In the presence of the inverted-terminal repeats (ITR), it forms a structure that is smaller and distinct from the structure formed with ssDNA and Rep68. Currently, we have cryo-EM data on the ssDNA/Rep68 complex and we have negative stain data on the ITR/Rep68 complex. All together, Rep68's flexibility appears to imply a more structural role for Rep68 to help hold the DNA in place as Rep40 pushes the DNA into the capsid.

82

POSTER NUMBER: 3012

Structure of the Mechanically Activated Ion Channel Piezo1

Kei Saotome[1], Swetha Murthy[1], Jennifer Kefauver[1], Ardem Patapoutian[1,2*], Andrew Ward[1*]
[1]The Scripps Research Institute, La Jolla, CA; [2]Howard Hughes Medical Institute
*Co-corresponding Author

Piezo1 and Piezo2 are mechanically activated ion channels that mediate touch perception, proprioception, and vascular development. Piezos are distinct from other ion channels and their structure remains poorly defined, impeding detailed study of their gating and ion permeation properties. Here, we report a high-resolution cryo-electron microscopy structure of the mouse Piezo1 trimer. The detergent-solubilized complex adopts a three-blade propeller shape with a curved transmembrane region containing at least 26 transmembrane helices per protomer. The flexible propeller blades can adopt distinct conformations, and consist of a series of four-transmembrane helix bundles we term 'Piezo Repeats'. Carboxy-terminal domains line the central ion pore, and the channel is closed by constrictions in the cytosol. The kinked alpha helical beam and anchor domain link the Piezo Repeats to the pore, and are poised to allosterically control gating. The structure provides a springboard to further dissect how Piezos are regulated by mechanical force.

83

POSTER NUMBER: 3013

Fibril structure of amyloid-ß(1-42) by cryo-electron microscopy

Lothar Gremer[1,2], Daniel Schölzel[1,2], Carla Schenk[1], Elke Reinartz[2], Jörg Labahn[1,2,3], Raimond B.G. Ravelli[4], Markus Tusche[1], Carmen Lopez-Iglesias[4], Wolfgang Hoyer[1,2], Henrike Heise[1,2], Dieter Willbold[1,2*], Gunnar F. Schröder[1,5*]
[1]Institute of Complex Systems, Structural Biochemistry (ICS-6), Forschungszentrum Jülich, 52425 Jülich, Germany; [2]Institut für Physikalische Biologie, Heinrich-Heine-Universität Düsseldorf, 40225 Düsseldorf, Germany; [3]Centre for Structural Systems Biology (CSSB), DESY, 22607 Hamburg, Germany Hamburg; [4]The Maastricht Multimodal Molecular Imaging Institute, Maastricht University, Universiteitssingel 50, 6229 ER Maastricht, The Netherlands; [5]Physics Department, Heinrich-Heine-Universität Düsseldorf, 40225 Düsseldorf, Germany.
*Corresponding Authors

Amyloids are associated with many diseases. Fibrillar aggregates of the amyloid-ß protein (Aß) are the main component of the senile plaques found in brains of Alzheimer's disease patients. We present the structure of an Aß(1-42) fibril determined by cryo-EM to a resolution of 4.0 Å. The presented polymorph of the fibril consists of two intertwined protofilaments in which individual subunits form a 'LS'-shaped topology. The backbone of all 42 residues and nearly all sidechains are well resolved in the EM density map, including the entire N-terminus, which is part of the cross-ß structure. The dimer interface protects the hydrophobic C-termini from the solvent. The unique staggering of the non-planar subunits results in markedly different fibril ends, termed "groove" and "ridge", leading to different binding pathways on both fibril ends, which has implications for fibril growth. The high-resolution structure is in agreement with solid-state NMR and X-ray diffraction data obtained from the same sample.

References:
Gremer et al., Science 358:116-119 (2017)

84

POSTER NUMBER: 3014

Cryo-EM structure of the synaptic wiring protein Teneurin

Moran Shalev-Benami[1], Jingxian Li[2], Demet Arac-Ozkan[2] and Georgios Skiniotis[1]
[1]Department of Molecular and Cellular Physiology & Department of Structural Biology, Stanford University;
[2]Department of Biochemistry and Molecular Biology, University of Chicago

Teneurins are highly conserved transmembrane receptors with large extracellular domains (ECDs) that play a crucial role in regulating synapse formation and organization in the central nervous system (CNS). Teneurin mutations and truncations have been previously linked to brain malformations, mental retardation, hyperactivity, schizophrenia and multiple types of cancers. Despite their central role, the exact function of teneurins in the CNS remains to be determined. Here we report the 3-Å cryo-electron microscopy structure of human teneurin ECD. The protein adopts an unusual architecture wherein a large cylindrical ß-barrel with striking similarity to bacterial toxins encapsulates a previously unidentified domain that highly resembles hormone and neuropeptide precursors. We demonstrate that this domain mediates interactions with additional protein partners at the synaptic cleft and dictate synapse formation. Coupled to biochemical and cell biology experiments, our results provide the first structural framework for delineating the unique and multifaceted functions of teneurins in synapse wiring.

85

POSTER NUMBER: 3015

Structure of the polycistic kidney disease TRP channel PC2

Chitra Shintre, Mariana Grieben, Ashley Pike, Elisa Venturi, Sam El-Ajouz, Annamaria Tessitore, Leela Shrestha, Shubhashish Mukhopadhyay, Pravin Mahajan, Rod Chalk, Nicola A Burgess-Brown, Rebecca Sitsapesan, Juha T Huiskonen & Elisabeth P Carpenter*
University of Oxford;
*Corresponding Author

Mutations in either polycystin-1 (PC1 or PKD1) or polycystin-2 (PC2, PKD2 or TRPP1) cause autosomal-dominant polycystic kidney disease (ADPKD) through unknown mechanisms. Here we present the structure of human PC2 in a closed conformation, solved by electron cryomicroscopy at 4.2-Å resolution. The structure reveals a novel polycystin-specific 'tetragonal opening for polycystins' (TOP) domain tightly bound to the top of a classic transient receptor potential (TRP) channel structure. The TOP domain is formed from two extensions to the voltage-sensor-like domain (VSLD); it covers the channel's endoplasmic reticulum lumen or extracellular surface and encloses an upper vestibule, above the pore filter, without blocking the ion-conduction pathway. The TOP-domain fold is conserved among the polycystins, including the homologous channel-like region of PC1, and is the site of a cluster of ADPKD-associated missense variants. Extensive contacts among the TOP-domain subunits, the pore and the VSLD provide ample scope for regulation through physical and chemical stimuli.

86

POSTER NUMBER: 3016

Molecular Dynamics Flexible Fitting at sub-5 Å Resolution

Abhishek Singharoy
Arizona State University, Biodesign Institute, Tempe, AZ, USA

High-resolution maps pose an imminent challenge to the traditional map-guided structure determination methods as the maps now characterize near-atomic scale features, the interpretation of which requires extremely precise structure building and validation protocols. Two structure determination methods, based on the molecular dynamics flexible fitting (MDFF) paradigm, are presented that resolve sub-5 Å cryo-electron microscopy (EM) maps with either single structures or ensembles of such structures. The methods, denoted cascade MDFF and resolution exchange MDFF, sequentially re-refine a search model against a series of maps of progressively higher resolutions, which ends with the original experimental resolution. Application of sequential re-refinement enables MDFF to achieve a radius of convergence of ~25 Å demonstrated with the accurate modeling of ß-galactosidase, TRPV1 and 26S proteasome at 3-4 Å resolution range. The MDFF refinements offer uniquely map-model uncertainty and B-factor determination criteria based on the inherent dynamics characterizing macromolecular ensembles, captured by means of local root mean square fluctuations. Finally, through combination with backbone-tracing algorithms and Maximum-Entropy methods, MDFF has been successfully extended to perform *ab initio* modeling of high-resolution maps.

87

POSTER NUMBER: 3036

Visualizing the Substrate Translocation Mechanism of the Hsp104 AAA+ Disaggregase

Stephanie Gates[1], Adam Yokom[1], Alexandrea Rizo[1], JiaBei Lin[2], Meredith E. Jackrel[2], James Shorter[2], and Daniel R. Southworth*[1,3]
[1]Life Sciences Institute, University of Michigan, Ann Arbor, MI
[2]Perelman School of Medicine, University of Pennsylvania, Philadelphia, PA
[3]Current Address: Dept. of Biochemistry and Biophysics, Institute for Neurodegenerative Diseases, University of California, San Francisco, CA

Heat shock protein (Hsp) 104 is a conserved member of the AAA+ family of hexameric-ring complexes that catalyze the translocation of polypeptide substrates through a central channel. Hsp104 functions as a disaggregase, re-solubilizing aggregated and toxic protein states in response to cell stress and to enhance prion propagation in yeast. The structural basis and conservation of the translocation mechanism remain fundamental questions across this broad family of AAA+ machines. Using cryo-electron microscopy, we have determined structures of active Hsp104 complexes to 4.0-Å resolution that identify a spiral of pore loop-substrate interactions spanning 80 Å along the channel and two distinct hexameric conformations that establish a ratchet-like translocation mechanism. Additional nucleotide-bound structures identify an open, "lock-washer" architecture we propose functions as a ground state for substrate engagement. 3D classification analysis under different conditions reveals these states are in equilibrium and substrate binding and ATP hydrolysis are required for conversion to the translocation-active state. Together, these results identify the structural basis for ATP-dependent polypeptide translocation by Hsp104 and reveal how processive and non-processive substrate-interaction mechanisms may function during dissaggregation.

88

POSTER NUMBER: 2036

Structural characterization of septin filaments and their membrane association

Sarah M. Sterling[1], Jeremy Thorner[1], Eva Nogales[1,2,3]
[1]Dept. of Molecular & Cell Biology, Univ. of California, Berkeley; [2]Howard Hughes Medical Institute; [3]Molecular Biophysics & Integrative Bio-Imaging Division, Lawrence Berkeley National Laboratory

Septins are a family of conserved GTP-binding proteins found in nearly all eukaryotes (except higher plants). In humans and budding yeast, four different septins self-associate into apolar hetero-octameric complexes. Depending on the terminal subunit, yeast hetero-octamers can polymerize end-on-end into long paired filaments or associate laterally to form spirals and rings. Septin-based structures serve as a scaffold for the recruitment of other proteins and have been demonstrated to form diffusion barriers and play roles in membrane remodeling. In yeast, septins make intimate contact with the plasma membrane via interaction with phosphatidylinositol-4,5-bisphosphate (PIP2). Our studies of the interaction of septin filaments with Bni5, a septin-binding protein that localizes to the bud neck during septin collar formation, indicated that Bni5 stabilizes and reorganizes septin filaments during cell cycle progression. Building on our mechanistic insights about Bni5-septin interaction, we are using Bni5 as a landmark for higher resolution studies of septin filaments utilizing cryo-EM. As an *in vitro* mimic of the interaction with plasma membranes, we are using PIP2-doped lipid monolayers as a substrate for cryo-EM characterization of Bni5-decorated septin filaments. Published crystal structures of truncated human septin monomers and full-length human septin complexes (reviewed in Ref. 1) revealed septin-septin contacts between the globular domain of these proteins, but the reported maps lack density for the C-terminal extension (CTE) of septins proposed to be important for filament pairing. Furthermore, given our evidence that the CTE of septin subunit Cdc11 is required for cooperative binding of Bni5 to septin filaments, a cryo-EM structure of septin filaments with Bni5 on a lipid monolayer will provide new insights about the Cdc11-Bni5 interaction, the septin filament-membrane interface, and the inter-filament contacts between paired septin filaments.

1. N. F. Valadares *et al.* (2017) Septin structure and filament assembly. *Biophys. Rev.* 9: 481-500.

Supported by NIH R01 Grants GM21841 (to J.T.) and GM101314 (to E.N. and J.T.).

89

POSTER NUMBER: 3018

Electron Cryo-Microscopy Reveals Essential Ebola Nucleoprotein Interactions in Nucleocapsid-like Helical Assembly

Zhaoming Su[1], Chao Wu[2], Grigore D. Pintilie[1], Muyuan Chen[1], Peng Ge[3], Gai Liu[2], Z. Hong Zhou[3], Christopher F. Basler[2], Daisy W. Leung[2], Gaya K. Amarasinghe[2,*], Wah Chiu[1,*]
[1]National Center for Macromolecular Imaging, Verna and Marrs McLean Department of Biochemistry and Molecular Biology, Baylor College of Medicine, Houston, Texas 77030 USA;
[2]Department of Pathology and Immunology, Washington University School of Medicine, St Louis, MO 63110 USA;
[3]Department of Microbiology, Immunology and Molecular Genetics, University of California, Los Angeles (UCLA), Los Angeles, California, USA
*To whom correspondence may be addressed.
Email: wah@bcm.edu, gamarasinghe@path.wustl.edu.

The Ebola virus nucleoproteins (eNPs) undergo self-assembly to form filamentous structures that serve as the scaffold of viral nucleocapsid (NC), which encapsidate the negative-sense viral RNA genomes in the presence of other viral proteins. The molecular determinants and mechanistic insights of NC assembly are critical therapeutic targets, yet has remained elusive. In the current study, we use electron cryo-microscopy (cryoEM) to determine the three-dimensional reconstruction of the eNP N-terminus domain 25-457 (ΔNP_{NTD}) that form nucleocapsid-like structures at ~6 Angstrom. The asymmetric unit of this left-handed filament is constituted of two ΔNP_{NTD} monomers. Docking of the ΔNP_{NTD} crystal structure reveals interactions that are essential to ΔNP_{NTD} self-assembly, which include the last two helices (21 and 22) in the foot lobe of ΔNP_{NTD} as the major oligomerization interfaces, as well as an extended helix in the NP N-terminal domain (385-411) that has not been previously resolved. Mutations in helices 21 and 22 disrupt NP oligomerization completely. This study provides the first detailed structural evidence of eNP nucleocapsid-like assembly and offers a framework towards understanding Ebola nucleocapsid architecture and remodeling during viral RNA replication.

90

POSTER NUMBER: 3017

Go Beyond Gctf in CTF Determination for Tilt Application in Single-Particle Cryo-EM

Min Su[1*]
[1]Life Sciences Institute, University of Michigan
*Corresponding Author (minsu@umich.edu)

Preferred specimen orientation is a common problem in single-particle Cryo-EM. Although adding detergent in the buffer has been shown effective to partially overcome such issue for certain sample, majorly of specimen suffer from preferred orientation to various degree when imbedded in ice, thus result in anisotropic map resolution. Single-particle Cryo-EM through tilt has been shown beneficial by utilizing gold grid for sample freezing. However, accurate CTF determination for each individual particle still remains as one of the major obstacles hinder tilt application in single-particle Cryo-EM. At the moment, Gctf serves as a compromised program has been applied to overcome such issue with limitations. Gctf was originally developed to refine particle defocus varies within the range of ice thickness for untitled images. Particularly, its weighting scheme in enhancing power spectrum by including neighboring particles is not suitable in the specimen tilt image acquisition. Here, we propose a revised weight scheme considering the geometry from a tilted micrograph. In addition, a pre-determined specimen tilt angle is calculated as a constrain to extrapolate defocus for each individual particle as an updated start defocus value for local refinement. This approach has been testified by experimental images. Moreover, it could potentially benifit for Cryo-ET in CTF local refinement.

References (optional)
Zhang. Gctf: Real-time CTF determination and correction. *J Struct Biol.* 2016.
Su *et al.* Mechanism of Vps4 hexamer function revealed by cryo-EM. *Sci Adv.* 2017.
Tan *et al.* Addressing preferred specimen orientation in single-particle cryo-EM through tilting. *Nat Methods.* 2017

91

POSTER NUMBER: 3019

Cryo-EM structure of alternative complex III/ *aa*3 cytochrome oxidase supercomplex from *Flavobacterium johnsoniae*

Chang Sun[1], Padmaja Venkatakrishnan[1], Samir Benlekbir[2], Yuhang Wang[1], John Rubinstein[2], Robert B. Gennis[1], Emad Tajkhorshid[1]
[1]University of Illinois Urbana-Champaign, Urbana, IL 61801, USA;
[2]The Hospital for Sick Children Research Institute, Toronto, Ontario M5G 0A4, Canada

Alternative complex III (ACIII) is an integral membrane protein recently identified in bacteria. It catalyzes the oxidoreduction between quinol and cytochrome *c*. It is a functional replacement for the canonical complex III of the respiratory electron transfer chain (e.g., the bc_1 complex). Bioinformatic studies have shown that the ACIII and the canonical complex III are evolutionarily unrelated and differ drastically in subunit and cofactor composition. There are proposals on how the ACIII functions but no structure is available. Here we present a structural study on the ACIII from *Flavobacterium johnsoniae* using single particle electron cryomicroscopy.
We prepared the protein sample with the styrene maleic acid (SMA) copolymer, which is an alternative to traditional detergents for membrane protein solubilization and purification. Analyses using SDS-PAGE, size exclusion chromatography and mass spectrometry showed that the ACIII forms a stable supercomplex with the aa_3-type cytochrome *c* oxidase. We prepared the specimen and acquired images at the Simons electron microscopy center.

From 160k particles, the density map of the supercomplex was determined to a resolution of 3.4 Å using cryoSPARC. A "divide-and-conquer" strategy was used to build the atomic structural model. The map was first segmented and studied to locate individual subunits. A model for each subunit was hand built in Coot. The assembled model was refined with both Phenix and molecular dynamics flexible fitting (MDFF). The final model of the ACIII consists of 6 subunits and more than 2,300 residues, covering 90% of the protein sequence. We were also able to model 10 lipid molecules that were clearly resolved in the ACIII density map, including 2 that are covalently linked to proteins, serving as anchors to the membrane bilayer. The stability of the ACIII model was confirmed in a 700-ns full-atom molecular dynamics simulation. The density quality for the cytochrome *aa*3 component of the supercomplex was not as good but allowed for a partial model, which revealed important protein-protein interactions at the ACIII/ *aa*3 interface.

92

POSTER NUMBER: 3020

Flagellum couples cell shape to motility in *Trypanosoma brucei*

Stella Y. Sun[1,2], Jason T. Kaelber[4], Muyuan Chen[4], Xiaoduo Dong[3], Jian Shi[3], Lim Chwee Teck[6], Steven J. Ludtke[4], Michael F. Schmid[1,2], Cynthia Y. He[3,5], Wah Chiu[1,2]
[1]Bioengineering Department, Stanford University; [2]SLAC National Accelerator Laboratory Bioscience Division; [3]Department of Biological Sciences, Center for BioImaging Sciences, National University of Singapore; [4]Verna and Marrs McLean Department of Biochemistry and Molecular Biology, Baylor College of Medicine; [5]Mechanobiology Institute, National University of Singapore; [6]Department of Mechanical Engineering, National University of Singapore

The flagellum is an integral component of motility in some cells, but its behavior has been rarely studied in relation to the cell body morphology. To understand how active beating of the flagellum couples cell shape to cell locomotion, we developed a miniaturized cell system using the single-celled parasite *Trypanosoma brucei* (*T. brucei*). We engineered two anucleate cell types that have different flagellar attachments to the cell body, and found these have different swimming behaviors. We used cryo-electron tomography (cryo-ET) to visualize these intact cells with biologically active flagella. Our videomicroscopy and cryo-ET analyses showed that flagellar motility is coupled to cytoskeleton deformation. Using microfluidic assays, we demonstrated that *T. brucei* can penetrate through an orifice smaller than its maximum diameter. Based on these observations, we propose a mechanism in which the force generated by the flagellum induces cell shape changes and cell body movement. This mechanism may be critical for *T. brucei* to infect its host through size-limiting barriers.

93

POSTER NUMBER: 3039

Differential Vaccinia Virus entry and innate immune signaling in macrophages

Siti Khadijah[1,2], Huei-Yin Cheng[1], Shu-Jung Chang[1], Chi-Hang Tsai[1] and Wen Chang[1,2*]
[1] Institute of Molecular Biology, Academia Sinica; [2]Graduate Institute of Life Sciences, National Defense Medical Centre,Taipei, Taiwan, Republic of China

Previously, we showed that Vaccinia virus Western Reserve (WR) contains A26 protein and enters HeLa cells through fluid phase endocytosis whereas a mutant, A26 deletion virus (WRA26) enters through plasma membrane fusion. Now we extended this finding into the primary innate immune cells particularly macrophages to understand whether WT-WR and WRA26 also enter immune cells through different pathways. Literatures showed that infection of WR in the human or mouse macrophages is not permissive, causing apoptosis of infected cells; however, the intracellular mechanism of restriction is unclear. We infected bone marrow derived macrophages (BMDM) with recombinant WT-WR and WRA26 expressing fluorescence early Venus and late mCherry marker proteins and analyzed viral gene expression kinetics using live imaging and flow cytometry. We found that WRA26 expressed robust early and late proteins while WT-WR only expressed limited amounts of early proteins in BMDM. Further experiments to understand the molecular basis that contributes to viral gene expression are currently underway and the results will be discussed.

94

POSTER NUMBER: 3021

Small angle X-ray scattering for data assisted protein structure prediction in CASP12 with prospects to improve accuracy of predictions

Tadeusz L. Ogorzalek[1], Greg L. Hura[1], Andriy Kryshtafovych[2], John A. Tainer[1,3], Krzysztof Fidelis[2], and Susan E. Tsutakawa[1*]
[1]Molecular Biophysics & Integrated Bioimaging, Lawrence Berkeley National Laboratory, Berkeley, California 94720, USA; [2]Protein Structure Prediction Center, Genome and Biomedical Sciences Facilities, University of California, Davis, CA 95616, USA; [3]Department of Molecular and Cellular Oncology, The University of Texas M. D. Anderson Cancer Center, Houston, Texas 77030, USA
*Presenter

Experimental data offers empowering constraints for structure prediction. These constraints can be used to filter equivalently scored models or more powerfully within optimization functions toward prediction. In CASP12, Small Angle X-ray Scattering (SAXS) data measured on an exemplary set of novel fold targets, were provided to the CASP community of protein structure predictors. As high throughput, solution-based techniques, SAXS can efficiently measure structural information on a protein in its native solution conformation and assembly. Here, we consider the SAXS data collected, the information in these solution measurements, and the challenges in incorporating them into computational prediction. As improvement opportunities were only partly realized in CASP12, we provide guidance on how data from the full-length biological unit and the solution state can better aid prediction of the folded monomer or subunit. We furthermore describe strategic integrations of solution measurements with computational prediction programs with the aim of substantially improving foundational knowledge and the accuracy of computational algorithms for biologically-relevant structure predictions for proteins in solution.

95

POSTER NUMBER: 2037

Visualizing polyclonal HIV vaccine responses using single particle electron microscopy

Hannah L. Turner[1], Matteo Bianchi[2], Bartek Nogel[1], Laura E. McCoy[2], Matthias Pauthner[2], Rebecca Nedellec[2], Katie McKenney[2], Dennis R. Burton[2], Lars Hangartner[2], and Andrew B. Ward[1]
[1]Department of Integrative Structural and Computational Biology, The Scripps Research Institute, USA; [2]Department of Immunology & Microbial Science, IAVI Neutralizing Antibody Center, The Scripps Research Institute, USA

An immune response against a pathogen or vaccine elicits a vast number of antibodies in an attempt to fend off disease. Isolating monoclonal antibodies (mAbs) for epitope mapping is an arduous process and is the rate-determining step in iterative structure-guided vaccine design. Traditionally, single particle EM has been used produce 3D models of antigen in complex with individual mAbs describing a single epitope. In this study, we describe a novel technique for epitope mapping an entire polyclonal response using a single EM dataset. Polyclonal antibodies from animals immunized with HIV vaccine candidates were purified, digested into Fabs, complexed with antigen, and imaged using either negative stain or cryoEM. The data were subjected to extensive 3D classification in Relion resulting in several individual Env trimer-Fab complexes that revealed the epitopes targeted by the polyclonal immune response. Processing samples from multiple animals taken at time points throughout an immunization study revealed antibody responses directed against an undesirable immunodominate epitope. From an immunized rabbit sample with 160k particles, we identified 5 distinct specificities of Fab complexed to antigen. A follow-up experiment using a 130k-particle dataset collected on a sample from an immunized Rhesus macaque identified a previously unknown epitope targeting the glycosylation site at N611 in the gp41 subunit. We also show that our methods are compatible with cryoEM and can generate high-resolution (<6 Å resolution) reconstructions of these polyclonal complexes. This new technique using EM to image the polyclonal immune response can quickly and accurately reveal the progression of antibody evolution directed towards multiple epitopes.

This work was supported by NIH grants HIVRAD P01 AI110657, and UM1 AI100663 (Scripps CHAVI-ID), the International AIDS Vaccine Initiative Neutralizing Antibody Center, and the Collaboration for Accelerating Vaccine Design OPP1115782 (Bill and Melinda Gates Foundation)

96

POSTER NUMBER: 3032

Structures of key intermediates in the human mitochondrial chaperonin reaction cycle

Yacob Gomez-Llorente[1*], Jebara Fady[2*], Shahar Nisemblat[2], Abdussalam Azem[2], Joel Hirsh[2], Iban Ubarretxena-Belandia[1]
[1]Department of Pharmacological Sciences, Icahn School of Medicine at Mount Sinai, New 2York, NY 10029, USA
[2]Department of Biochemistry and Molecular Biology, The George S. Wise Faculty of Life Sciences, Tel Aviv University, Tel Aviv, Israel.
Corresponding: iban.ubarretxena@gmail.com & jhirsch@post.tau.ac.il
* Equally contributed to this work

The human mitochondrial chaperonin, mHsp60, together with its co-chaperonin mHsp10, mediates the folding of newly imported mitochondrial proteins, and corrects misfolded polypeptides resulting from mitochondrial stress in the matrix. We will present a 3.0 Å resolution cryo-EM 3-D reconstruction of a mHsp60:mHsp10 football complex. All the fourteen nucleotide-binding sites in the complex are occupied by ADP and Magnesium, indicating the complex was solved in a post ATP-hydrolysis state. With a unique contact interface between two mHsp60 rings our complex represents a late-stage dissociation intermediate in the reaction cycle of the chaperonin.

97

POSTER NUMBER: 3023

A Point Mutation that Leads to a Human Neurodegenerative Disorder Destabilizes the Heat Shock Protein 60 Chaperonin

Jinliang Wang, Ricardo A. Bernal
The University of Texas at El Paso

Protein folding is a critical process that prevents the loss of protein function and the accumulation of aggregated misfolded proteins. Chaperonins are large protein complexes that have evolved to aid in protein folding of newly synthesized that cannot fold on their own or proteins that have become partially misfolded. Hsp60 is a human chaperonin that has been linked to human disease. Hereditary spastic paraplegia (HSP) is a neurodegenerative disorder that belongs to a larger group of inherited conditions characterized by stiffness (spasticity) and a slow, progressive weakening of the lower limbs. Although it is slow to develop, people with the disease will eventually require assistance in walking and may become confined to a wheelchair. The mutation (V72I) in the HSPD1 gene results in severe psychomotor developmental delay, involuntary eye movement (nystagmus), low muscle tone (hypotonia) and weakness, and prominent stiffness (spasticity). Studies have found that an hsp60 mutation (V72I) results in diminished protein folding activity and is likely the cause of hereditary spastic paraplegia (SPG13). Here we will present an hsp60 V72I mutant reconstruction reveals details about the cause of hereditary spastic paraplegia. The 9-angstrom resolution reconstruction is composed of 14 identical subunits arranged into a closed double ring. Hsp10, a co-chaperonin that works in conjunction with the wild-type complex, does not bind to the V72I mutant. Furthermore, the V72I hsp60 mutation reduced the ability of the chaperonin to refold denatured protein by almost 5-fold *in-vitro*. It appears that binding of nucleotide destabilizes the V72I chaperonin complex which in turn negatively affects its protein folding ability and leads to the disease.

Keywords: hsp60 V72I, EM structure, hereditary spastic paraplegia

98

POSTER NUMBER: 3024

Multisegment Molecular Dynamics Flexible Fitting for cryo-EM

Yuhang Wang and Emad Tajkhorshid
University of Illinois, Urbana-Champaign, Urbana, IL, USA

Molecular Dynamics Flexible Fitting (MDFF) has become a popular method for determining molecular structures by fitting existing structural models into the experimental single-particle cryo-EM density maps. The Recent extensions of the MDFF, namely cascade MDFF (cMDFF) and resolution exchange MDFF (ReMDFF), bring faster convergence to the structural fitting process. However, the current MDFF techniques do not distinguish the regions of cryo-EM map of different resolutions and sometimes lead to underperformance. Here we demonstrate a generalized MDFF methodology named "Multisegment Molecular Dynamics Flexible Fitting" (segMDFF). It creates customized fitting scheme for individual regions of the target protein. In the replica exchange framework, different regions experience exchanges independently, creating a free recombination of local maps of different resolutions. In addition to the resolution exchange, we also explore additional reaction coordinates such as temperature and grid coupling scaling factor.

99

POSTER NUMBER: 3025

The *in situ* structural analysis of a pathogenic mutant LRRK2 protein in Parkinson's Disease using cryo-electron tomography

Watanabe R.[1], Buschauer R.[1,2], Böhning J.[1], Boassa D.[3], Taylor S.[4], and Villa E.[1]
[1]Department of Molecular Biology, University of California San Diego; [2]Present address Gene Center of the University of Munich; [3]The National Center for Microscopy and Imaging Research, University of California San Diego; [4]Department of Chemistry & Biochemistry, Department of Pharmacology, University of California San Diego

Mutations in the leucine-rich repeat kinase 2 (LRRK2) are the major cause for familial Parkinson's disease (PD). LRRK2 encodes a large protein of 2527 amino acids composed of kinase and GTPase domains surrounded by multiple protein-protein interaction regions. Most of the pathological mutations are found in the catalytic domains, suggesting modulation of GTPase and kinase activities may play a crucial role in pathogenesis. Although being a potentially promising drug target, full-length LRRK2 has eluded structural determination. Furthermore, while numerous studies showed that various pathological mutations or loss of LRRK2 gene affect various cellular processes including cytoskeleton and organelle integrity, secretory and endocytic pathways, autophagy and mitochondrial homeostasis, the precise function of LRRK2 at the molecular level remains elusive.
Various pathological mutants of LRRK2 have been shown to form filamentous structures surrounding microtubules. This phenotype has been observed not only with mutant LRRK2 proteins but also with wild-type LRRK2 after pharmacological LRRK2 kinase inhibition or by addition of a non-hydrolyzable GTP analogue. This suggests that LRRK2 is arrested at a certain conformation on microtubules, which could represent the pathogenic form of LRRK2 protein in general, emphasizing the need of detailed structural information for development of future PD therapeutics.

Cryo-electron tomography (CET) can directly unveil the structure of biomolecules *in situ* at molecular resolution in 3-D. However, the applicability of CET is limited to samples less than 500-nm in thickness, which is exceeded by eukaryotic cells. In order to reveal the structure of a pathological mutant of LRRK2 (I2020T) decorating microtubules in cells, we used cryo-correlative light and electron microscopy (CLEM) and cryo-focused ion beam (cryo-FIB) milling to prepare samples for cryo-electron tomography. CLEM allows us to target the specific region of cells containing filamentous structures of YFP-tagged LRRK2 to be thinned using focused ion beam milling. Subsequently these thin regions of cells supported by the rest of cells were imaged by cryo electron tomography. We found that microtubules decorated with LRRK2 form bundle-like structures with regular spacing. Interestingly, in contrast to the canonical microtubules composed of *13 protofilaments* found in mammalian cells, LRRK2 (I2020T) decorated microtubules are composed of 12 protofilaments. The structure of LRRK2 decorating microtubules is described in detail in an accompanying presentation from our laboratory.

Our results suggest that mutant LRRK2 might alter the cellular microtubule structure, which likely affects its dynamics and/or roles for various cellular processes mediated by microtubules and ultimately leads to the pathological phenotype found in PD patients.

100

POSTER NUMBER: 3026

Visualizing mechanisms of influenza virus neutralization by intact IgG and Fab fragments requires complementary electron microscopy approaches

James Williams, Long Gui, Alexander Mileant, Kelly K. Lee
University of Washington

The neutralizing antibody (nAb) response against the influenza virus's hemagglutinin (HA) fusion glycoprotein is important for preventing viral infection, but we lack a complete understanding of the mechanisms by which nAbs act. Available crystal structures of antigen binding fragments (Fab) in complex with soluble isolated antigen constructs have helped to generate hypothetical models of neutralization. Based on such structures, it is usually inferred that blockage of receptor binding or inhibition of conformational changes needed to carry out fusion are responsible for neutralization. Interactions between IgG and whole virus however, have been much less characterized.

To understand mechanisms of IgG-mediated neutralization requires a multiscale approach that enables one to visualize isolated Fab-antigen complexes as well as whole IgG interacting with intact virus. We used a combination of biophysical approaches, single particle EM, and cryo-electron tomography (cryo-ET) to investigate the effect of nAb binding and the role of IgG bivalency on inhibition of HA function for nAbs targeting different epitopes on the HA spike.

HC19 binds the receptor-binding pocket at HA's distal head while FI6v3 binds the HA2 fusion subunit. Single particle EM showed that nAbs such as HC19, previously thought to act by blocking receptor binding, also inhibit late-stage HA conformational changes required for fusion. Stem-directed Abs such as FI6v3 also stabilized a prefusion form of HA. Cryo-ET demonstrated that IgG bivalency, notably for HC19, added an additional mode of inhibition by crosslinking virus particles leading to formation of aggregates in which large proportions of HA were buried and hence blocked from mediating fusion. The stem-targeting nAb FI6v3 did not aggregate particles. Furthermore, monovalent FI6v3 Fab was significantly less effective than IgG in inhibiting fusion activity. We deduced that *interspike* crosslinking within a given particle by FI6v3 IgG may disrupt spike mobility on the virus surface, which is needed to drive fusion. EM is thus an ideal technique that enables characterization of both the local impact of Fab binding as well as the global effect of IgG bivalency on mechanisms of virus neutralization.

Funding: NIH R01-GM099989

101

POSTER NUMBER: 3033

Assembly of the Bacterial Large Ribosomal Subunit

Jessica Rabuck-Gibbons[1], Joseph P. Davis[2], Yong-Zi Tan[4], Carla Cervantes[1], Emily Green[1], Philip Baldwin[3], Clint Potter[4], Bridget Carragher[4], Dmitry Lyumkis[3], James R. Williamson[1*],
[1]The Scripps Research Institute; [2]Massachusetts Institute of Technology, [3]New York Structural Biology Center, [4] The Salk Institute.

The large, or 50S, ribosomal subunit (LSU) is a 1.3 MDa ribonucleoprotein complex formed from two large ribosomal RNAs and 35 small proteins, forming the catalytic subunit of the full 70S ribosome. Assembly of the LSU occurs rapidly in cells, and the transcription, RNA processing, RNA folding, RNA modification, and protein binding required for LSU formation is complete within 3-4 minutes. Efficient assembly under a wide variety of conditions is facilitated by over 20 specific protein factors that are helicases, GTPs and chaperones. We are investigating the structures of LSU assembly intermediates that accumulate when the assembly pathway is perturbed, in order to determine the parallel and sequential elements of the LSU assembly. We use a combined quantitative mass spectrometry and electron microscopy approach to analyze the composition and structure of the heterogeneous sets of intermediates that accumulate during assembly. By limitation of expression of the essential ribosomal protein L17, a set of four major intermediates was observed, that could be sub-classified into additional species. These species could be placed into a putative assembly order, allowing proposal of a specific LSU assembly mechanism for the first time. In addition, cooperative RNA folding blocks were identified that correspond folding units that span multiple domains of RNA secondary structure. Additional results will be presented on the impacts of the DEAD-box helicase SrmB on LSU assembly, as well as new particles observed upon limitation of expression of additional ribosomal proteins. We are working toward a comprehensive mechanistic understanding of ribosome assembly in bacteria.

102

POSTER NUMBER: 3027

Using X-ray Free Electron lasers to visualize solvent in the M2 proton channel

Rahel A. Woldeyes[1], Jessica L. Thomaston[2], Takanori Nakane[3], Aaron S. Brewster[4], Tomoyuki Tanaka[5], Monarin Uervirojnangkoorn[6], Benjamin A. Barad[1], Ayumi Yamashita[5], Yujie Chen[7], Toshi Arima[5], Jun Kobayashi[5], Tetsuya Masuda[8], Mamoru Suzuki[9], Fumiaki Yumoto[8], Michihiro Sugahara[5], Nicholas K. Sauter[4], Rie Tanaka[5], Eriko Nango[5], So Iwata[5], William F. DeGrado[2], James S. Fraser[1]
[1]Department of Bioengineering and Therapeutic Sciences, [2]Department of Pharmaceutical Chemistry, University of California- San Francisco; [3]The University of Tokyo; [4]Physical Biosciences Division, Lawrence Berkeley National Laboratory; [5]RIKEN SPring-8 Center; [6]Stanford University; [7]Cornell University; [8]Kyoto University; [9]Osaka University; [10]High Energy Accelerator Research Organization, KEK

The influenza A virus must acidify its interior to release its RNA. This important step in the viral life cycle relies on matrix protein 2 (M2), a pH-regulated proton channel embedded in the lipid envelope of the virus. M2 has been a subject of intense study both because it the target of adamantane compounds that can be used to treat flu and because it is an ideal model system to study proton conduction across membranes. The structural location of ordered water within the M2 channel plays a key role in both of these areas. Ordered water molecules are proposed to make key bridging interactions between adamantane compounds and the protein. Additionally, there is significant debate about the existence "water wires" of ordered solvent molecules that allow proton conduction by connecting the viral exterior and interior. It is therefore critical to determine the distribution of water molecules within the channel. Previous high resolution X-ray data of the M2 channel crystallized in the lipid sponge phase revealed interesting differences in the distributions of water as a function of pH. However, preliminary room temperature data called into question whether data collected from cryo-cooled crystals was informative for biologically relevant mechanisms of proton transport. Unfortunately, damage to the small M2 crystals from the radiation dose at room temperature precluded mechanistic interpretations of the solvent density. To circumvent these issues and to observe the functionally important structural dynamics of M2, we used X-ray Free Electron Lasers to collect radiation-damage free, 1.4 Å dataset of M2 at low and high pH. By avoiding the artifacts associated with cryo-cooling and eliminating the effects of radiation damage, the resulting room temperature data have allowed us to draw new conclusions about the solvent conformational heterogeneity. We will build on this work and use XFEL to monitor changes in the solvation of the pore of M2 as a function of both pH and time. Furthermore, these structures will help in the design of next-generation of antivirals address continuing problems with drug resistance.

103

POSTER NUMBER: 3037

CryoEM structure and biochemical analysis reveal assembly and regulation mechanisms of Ebola nucleocapsid in viral RNA synthesis

Chao Wu[1], Zhaoming Su[2], Liuqing Shi[3], Priya Luthra[4], Grigore D. Pintilie[5], Britney Johnson[1], Justin R. Porter[6], Peng Ge[7], Muyuan Chen[5], Gai Liu[1], Thomas E. Frederick[6], Jennifer M. Binning[1], Gregory R. Bowman[5], Z. Hong Zhou[7], Christopher F. Basler[4], Michael L. Gross[3], Daisy W. Leung[1], Wah Chiu[2*], and Gaya K. Amarasinghe[1*]
[1]Department of Pathology and Immunology, Washington University School of Medicine [2]Department of Bioengineering and Department of Microbiology and Immunology, James H. Clark Center, Stanford University
[3]Department of Chemistry, Washington University in St. Louis
[4]Center for Microbial Pathogenesis, Institute for Biomedical Sciences, Georgia State University
[5]Verna and Marrs McLean Department of Biochemistry and Molecular Biology, Baylor College of Medicine
[6]Department of Biochemistry and Molecular Biophysics, Washington University School of Medicine
[7]Department of Microbiology, Immunology and Molecular Genetics, University of California, Los Angeles (UCLA)
*Corresponding Author

Ebola virus (EBOV) causes severe hemorrhagic fever with high case fatality rates during outbreaks. EBOV viral genome is encapsidated by nucleoproteins (eNPs) which self-assemble into tubular structures. These structures act as scaffolds of nucleocapsids (NCs) which serve as a template for viral RNA synthesis. Molecular insights into assembly and regulation mechanisms of nucleocapsid are useful for developing specific treatments targeting viral RNA synthesis. Here we describe biochemical characterizations of eNP N-terminal domain (NTD) nucleocapsid-like structures and insights gained from a 5.8 angstrom cryoEM structure. Hydrogen-deuterium exchange mass spectrometry and biochemical data indicate that α-helices 21 and 22 from the foot lobe form coiled-coil contacts that are essential for nucleocapsid-like assembly and ssRNA binding. Using a monomeric mutant, which contains mutations at this interface, we found that the eNP N-terminus (residues 1-24) also contributes to eNP oligomerization. Its presence confers eNP with high affinity ssRNA binding while reducing affinity for eVP35 NPBP, a region involved in chaperoning eNP in an RNA-free state. In addition, in our cryoEM structure, an additional α-helix 23 occupies the eVP35 NPBP pocket and this region is critical for minigenome activity. Taken together, our study identifies both the eNP N-terminus and α-helices 21-23 as important regulatory elements and offers a framework towards understanding Ebola NCs structure and remodeling during viral RNA synthesis.

104

POSTER NUMBER: 3034

Native MS in Structural Biology:
Surface Collisions of Protein and Nucleoprotein Complexes

Vicki Wysocki[1*], Yang Song[1], Vahe Bandaria[2], Daniel Nemecek[3]
[1]The Ohio State University; [2]University of Utah, [3]FEI/ThermoFisher; Corresponding author*

Structural biology methods such as cyroEM and nuclear magnetic resonance provide high-resolution information on the structures of protein complexes. Native mass spectrometry is a complementary approach that provides critical structural information with higher throughput on low sample amounts. The power of native MS increases when coupled to ion mobility (IM-MS), a technique that measures rotationally averaged collisional cross sections and thus direct information on conformational changes, or to high resolution mass spectrometry (HRMS). This presentation will illustrate surface-induced dissociation/ion mobility SID/IM MS and SID HRMS for characterization of topology, intersubunit connectivity, and other structural features of multimeric protein complexes. Data for a number of protein-partner complexes are under investigation, where the partner can be small molecule ligand, protein, DNA, or RNA. We are also optimizing online separations approaches coupled to the native MS/SID/IM experiments. For toyocamycin nitrile hydratase, we proposed a dimer-of-trimers heterohexameric structure based on multiple MS methods and computations. A cryoEM structure determination has been started for comparison with the native MS results and the first images show an elongated structure of about 13 nm in length and about 8 nm in diameter.

105

POSTER NUMBER: 3038

Structural insights into the role of phosphorylated IRF3 in IFN-β transcription

Pengbiao Xu[1], Tao Jing[1], Baoyu Zhao[1], Fenglei Du[1], Pingwei Li[1*],
[1]Texas A&M University
*Corresponding Author

Cells can defend against bacteria and viruses' invasion through their immune system. The recognition of pathogen-derived dsDNA or dsRNA is a strategy to initiate potent immune response by the host. Previous studies showed that cGAS-STING signaling pathway is involved in the detection of dsDNA, which ultimately initiates the induction of type I interferons (IFNs) through adaptor protein STING and transcription factor IRF3. Previous studies indicated that upon infection, IRF-3 in the host cell is activated through phosphorylation at the C-terminal SRR region to form dimers. The pIRF-3 dimerization leads to its translocation to the nucleus followed by binding IFN-β promoter to induce the transcription of IFN-β. So far there have been extensive studies focusing on the transcriptional initiation of the IFN-β gene through an IFN-β enhanceosome that assembles IRF3. However, how the full-length phosphorylated IRF3 binds to the full-length enhancer and then subsequently initiates the transcription of IFN-β remains elusive. To elucidate the molecular mechanism of IFN-β transcription by the phosphorylated IRF3, we are working on determining the structure of full-length phosphorylated IRF-3 in complex with dsDNA derived from the IFN-β enhancer by Cryo-EM. The full-length pIRF3 mixed with dsDNA is purified through size-exclusion column and applied to negative stain analysis, and subsequent Cryo-EM data collection is in progress.

References
1. Paludan SR, Bowie AG: Immune sensing of DNA. Immunity 2013, 38:870-880.
2.Shu C, Li X, Li P: The mechanism of double-stranded DNA sensing through the cGAS-STING pathway. Cytokine Growth Factor Rev 2014, 25:641-648
3. Panne, D., Maniatis, T., and Harrison, S.C. (2004). Crystal structure of ATF-2/c-Jun and IRF-3 bound to the interferon-beta enhancer. EMBO J. 23, 4384-4393

Funding
1.The Robert Welch Foundation. Grant A-1816
2.CPRIT grant.

106

POSTER NUMBER: 3028

Conformational dynamics reveal translocation mechanism of Hsp104, a yeast AAA+ disaggregase

Adam Yokom[1,2], Stephanie Gates[1,2], JiaBei Lin[3], Meredith E. Jackrel[3], Alexandrea N. Rizo[1], James Shorter[3], Daniel R. Southworth[1,4*]
[1]University of Michigan, Life Science Institute, [2]Current address : University of Berkeley, Department of Molecular and Cell Biology, [3] University of Pennsylvania, Department of Biochemistry and Biophysics, [4]Current Address: University of San Francisco, Institute of Neurodegenerative Disease
*Corresponding Author: daniel.southworth@ucsf.edu

AAA+ (ATPases Associated with diverse cellular Activities) protein machines have long been a focus of structural biology research due to their ubiquitous cellular involvement including DNA replication, protein degradation and disaggregation. Understanding how one family of proteins can perform a wide range of mechanistic processes has proven difficult as the AAA+ proteins differ in functional oligomeric state, ATPase activity and are thought to be dynamic in nature. Current structural information has failed to explain the mechanistic underpinnings of the translocation mechanism of AAA+ proteins. To investigate this mechanism, structural and biochemical studies were conducted on Hsp104, a AAA+ disaggregase found in yeast. Hsp104 natively aids in prion propagation and solubilizing amorphous aggregates but has been shown to disassemble human amyloids as well. Using cryo-EM we captured a pair of high resolution reconstructions of Hsp104 bound with substrate and ATP-gamma-S. These structures reveal two distinct conformational states which show substrate-pore loop interactions, a ratchet-like translocation step and a processive rotary mechanism for polypeptide disaggregation. Additionally, this study highlights the power of conformational sorting *in silico* to capture the plasticity of dynamic protein machines.

Gates SN*, Yokom AL*, Lin J, et al. Ratchet-like polypeptide translocation mechanism of the AAA+ disaggregase Hsp104. Science. 2017;357(6348):273-279. *Contributed equally

107

POSTER NUMBER: 3035

Ribosome labeling: a method for high-resolution structural analysis of small RNAs using cryoEM

Cheng Zhang[1], William Cantara[2], Karin Musier-Forsyth[2], Dmitry Lyumkis[1]
[1]Laboratory of Genetics and Helmsley Center for Genomic Medicine, Salk Institute for Biological Studies
[2]Department of Chemistry and Biochemistry, Ohio State University

Recent advances in single particle cryo-electron microscopy (cryoEM) has enabled structure determination of proteins and macromolecular complexes with sizes larger than 200 kDa to near-atomic resolution (< 4Å). However, it is still challenging for high resolution structure determination of small proteins (<100 kDa) and small RNAs (< 100 bp) by cryoEM. Here we developed a method called ribosome labeling that adds mass to small RNAs by fusing them to the helix of 23S rRNA of E. Coli and the chimeric ribosomes are then purified from SQZ10 (Δ 7 prrn strain)[1] with ~100% occupancy. Combined with focused classification, we are able to deconvolute RNA mobility at the expected region and obtain near-atomic resolution. In order to find out the most stable region on the ribosome, we quantitatively compared HIV-1 TAR RNA insertions at ten helices of 23s rRNA. Our results showed that helix 68 was the most promising candidate to fuse RNA of interest. Moreover, ribosome labeling is applicable for solving small RNP structures.

References
1. Daoming Qin et al., RNA (2007)
2. Jonathan Pollack et al., J Virol (1993)

108

POSTER NUMBER: 3029

CryoEM structure of the yeast spliceosomal P complex at 3.3 Å resolution, the last major piece of the missing puzzle in the splicing cycle

Rui Zhao[1], Shiheng Liu[2,3], Xueni Li[1], Lingdi Zhang[1], Jiansen Jiang[2,3], Ryan C. Hill[1], Yanxiang Cui[2], Kirk C. Hansen[1], Z. Hong Zhou[2,3]
[1]Department of Biochemistry and Molecular Genetics, University of Colorado Denver Anschutz Medical Campus, Aurora, CO 80045, USA; [2]Electron Imaging Center for Nanomachines University of California, Los Angeles (UCLA), Los Angeles, CA 90095, USA; [3]Department of Microbiology, Immunology, and Molecular Genetics, UCLA, Los Angeles, CA 90095, USA

The spliceosome undergoes dramatic changes signified by the B, B^{act}, B*, C, C*, P, and ILS complexes in a splicing cycle. Recent cryoEM structures of B, B^{act}, C, C*, and ILS revealed mechanisms of 5' ss recognition, branching, and intron release, but lacked information on 3' ss recognition, exon ligation and release, all related to the post-catalytic P complex. Here, we report the cryoEM structure of the yeast P complex at 3.3Å resolution, the highest among all spliceosomal structures reported to date. Our atomic model of the P complex reveals that the 3' ss recognition is driven by the interaction between 3' and 5' ss, likely facilitated by a stem-like structure formed between the intronic regions close to the branch site and 3' ss. Our structure reveals that one or more new proteins become stably associated with the core components of the P complex around Prp8 and Prp22, securing the 3' exon and potentially regulating the activity of Prp22. The structure demonstrates that Prp22 binds 15-21 nucleotides downstream of the exon-exon junction, enabling it to pull the 3' intron-exon and ligated exon in a 3' to 5' direction to achieve 3' ss proofreading and exon release, respectively. Thus, by providing insights into the molecular mechanism of 3' ss recognition, exon ligation, and the action of RNA helicase Prp22, the atomic model of the P complex fills the last major gap in our understanding of the splicing cycle.

109

POSTER NUMBER: 3030

Single particle cryoEM reveals the specific recognition of centromeric nucleosomes by CENP-N

Keda Zhou (1), Satyakrishna Pentakota (2), Garry P. Morgan (3), Ingrid R. Vetter (2), Andrea Musacchio (2) & Karolin Luger (1,4)
(1) Department of Chemistry and Biochemistry, University of Colorado at Boulder; 3415 Colorado Avenue, Boulder, CO - 80303
(2) Department of Mechanistic Cell Biology, Max-Planck Institute of Molecular Physiology, Otto-Hahn-Straße 11, 44227 Dortmund, Germany
(3) Department of MCDB, University of Colorado at Boulder; 3415 Colorado Avenue, Boulder, CO - 80303
(4) Howard Hughes Medical Institute

Centromere protein (CENP) A is a histone H3 variant and a key epigenetic determinant of a specialized chromosome domains known as centromeres. These nucleate kinetochores[1], multi-subunit complexes that capture spindle microtubules to promote chromosome segregation during mitosis. CENP-A localizes to centromeres in a DNA-sequence-independent manner. Like H3, it is incorporated into nucleosomes together with histones H2A, H2B, and H4 to from CENP-A nucleosomes that are interspersed with H3-containing nucleosomes. Two kinetochore proteins, CENP-C and CENP-N, directly recognize CENP-A in the context of the nucleosome. How these proteins 'ignore' the abundant H3 nucleosomes to interact selectively with a handful of centromeric CENP-A nucleosomes has important implications for genome stability during cell division. By single particle cryo electron microscopy, we obtained a 4.0 Å density map of the CENP-N:CENP-A nucleosome complex. By fitting the known crystal structure of CENP-A nucleosome together with the newly determined crystal structure of CENP-N, we reveal the structural basis on how CENP-N specifically recognizes the nucleosome containing CENP-A. Through charge and space complementarity, CENP-N decodes the L1 loop that is unique to CENP-A. CENP-N also engages in extensive interactions with a long segment of the distorted nucleosomal DNA double helix. We validate these interactions *in vitro* and *in vivo*. Unexpectedly, the orientation of H4 N-terminal tail is altered due to its interaction with CENP-N, with important implications for the multiple biological processes involving H4 N-terminus. Collectively, our studies shed light on a fundamental molecular mechanism at centromere.

110

POSTER NUMBER: 3031

The Primed SNARE-Complexin-Synaptotagmin Complex for Neurotransmitter Release

Qiangjun Zhou[1,2], Peng Zhou[1], Thomas C. Südhof[1] & Axel T. Brunger[1,2]
[1]Department of Molecular and Cellular Physiology, Howard Hughes Medical Institute, Stanford University, Stanford, California 94305, USA; [2]Department of Neurology and Neurological Sciences, Department of Structural Biology, Department of Photon Science, Stanford University, Stanford, California 94305, USA

Synaptic transmission is essential for the process of communication between two neurons. It occurs upon fusion of synaptic vesicles with the plasma membrane, a highly regulated and ultrafast process occurring in less than a millisecond. Synaptotagmin, complexin, and neuronal SNARE (soluble *N*-ethylmaleimide sensitive factor attachment protein receptor) proteins mediate evoked synchronous neurotransmitter release, but the molecular mechanisms mediating the cooperation between these molecules remain unclear. We determined crystal structures of the SNARE-synaptotagmin-1 complex (Zhou Q., *Nature* 2015) and the primed pre-fusion SNARE-complexin-synaptotagmin-1 complex (Zhou Q., *Nature* 2017). These structures reveal a specific, Ca^{2+}-independent, conserved interface between synaptotagmin-1 and the SNARE complex. The structure of the primed pre-fusion SNARE-complexin-synaptotagmin-1 complex reveals an unexpected tripartite interface between synaptotagmin-1 and both the SNARE complex and complexin. Mutations that disrupt either interface in solution also severely impair evoked synchronous release in neurons, suggesting that both interfaces are essential for the primed pre-fusion state. Ca^{2+} binding to the synaptotagmin-1 molecules unlocks the complex, allows full zippering of the

SNARE complex, and triggers membrane fusion. The tripartite SNARE-complexin-synaptotagmin-1 complex at a synaptic vesicle docking site has to be unlocked for triggered fusion to start, explaining the cooperation between complexin and synaptotagmin-1 in synchronizing evoked release on the sub-millisecond timescale (Zhou Q., *Nature* 2015, 2017). However, the molecular organization of the primed pre-fusion complexes, and the association of these complexes with the synaptic and plasma membranes remains unclear. Now I am working on the structure determination of the primed pre-fusion complex in a "native" membrane environment using single-particle cryo-electron microscopy (cryo-EM). Moreover, I am also studying the molecular organization of the priming state of synaptic vesicles and plasma membrane in cultured neurons using *in situ* cellular cryo-electron tomography (cryo-ET).

References:

1. Zhou Q. *et al*, (2017) The Primed SNARE-Complexin-Synaptotagmin Complex for Neuronal Exocytosis. *Nature,* 548:420-425 (Featured on the cover)

2. Zhou Q. *et al*, (2015) Architecture of the Synaptotagmin-SNARE Machinery for Neuronal Exocytosis. *Nature,* 525:62-67

Funding: Pathway to Independence Award (K99/R00, 1K99MH113764-01), National Institute of Mental Health

111

Keystone Symposia
PO Box 1630
Silverthorne, CO 80498
USA

ISBN 978-1-5108-6255-5

9 781510 862555

CURRAN ASSOCIATES INC.
proceedings.com